GETTYSBURG COLLEGE
LIBRARY

GETTYSBURG, PA

GRASSLAND GROUSE

Also by Paul A. Johnsgard

The Nature of Nebraska: Ecology and Biodiversity (2001)
Prairie Birds: Fragile Splendor in the Great Plains (2001)
Trogons and Quetzals of the World (2000)
The Pheasants of the World: Biology and Natural History (2nd ed. 1999; 1st ed. 1986)
Earth, Water, and Sky: A Naturalist's Stories and Sketches (1999)
Baby Bird Portraits by George Miksch Sutton: Watercolors in the Field Museum (1998)
The Avian Brood Parasites: Deception at the Nest (1997)
The Hummingbirds of North America (2nd ed. 1997; 1st ed. 1983)
Ruddy Ducks and Other Stifftails: Their Behavior and Biology (with M. Carbonell) (1996)
This Fragile Land: A Natural History of the Nebraska Sandhills (1995)
Arena Birds: Sexual Selection and Behavior (1994)
Cormorants, Darters, and Pelicans of the World (1993)
Ducks in the Wild: Conserving Waterfowl and Their Habitats (1992)
Bustards, Hemipodes, and Sandgrouse: Birds of Dry Places (1992)
Crane Music: The North American Cranes (1991)
Hawks, Eagles, and Falcons of North America: Biology and Natural History (1990)
Waterfowl of North America: The Complete Ducks, Geese, and Swans (with S. D. Ripley and R. Hill) (1989)
North American Owls: Biology and Natural History (1988)
The Quails, Partridges, and Francolins of the World (1988)
Diving Birds of North America (1987)
Birds of the Rocky Mountains (1986)
Prairie Children, Mountain Dreams (1985)
The Platte: Channels in Time (1984)
The Cranes of the World (1983)
The Grouse of the World (1983)
Dragons and Unicorns: A Natural History (with Karin Johnsgard) (1982)
Teton Wildlife: Observations by a Naturalist (1982)
Those of the Gray Wind: The Sandhill Cranes (1981)
The Plovers, Sandpipers, and Snipes of the World (1981)
A Guide to North American Waterfowl (1979)
Birds of the Great Plains: Breeding Species and Their Distribution (1979)
Ducks, Geese, and Swans of the World (1978)
The Bird Decoy: An American Art Form (editor) (1976)
Waterfowl of North America (1975)
American Game Birds of Upland and Shoreline (1975)
Song of the North Wind: A Story of the Snow Goose (1974)
Grouse and Quails of North America (1973)
Animal Behavior (2nd ed. 1972; 1st ed. 1967)
Waterfowl: Their Biology and Natural History (1968)
Handbook of Waterfowl Behavior (1965)

GRASSLAND GROUSE
and Their Conservation

PAUL A. JOHNSGARD

SMITHSONIAN INSTITUTION PRESS
Washington and London

Dedicated to the prairie grouse pioneers: Alfred Gross,
Aldo Leopold, Ingemar Hjorth, Frederick and
Frances Hamerstrom, and, of course, all the Gaboons

© 2002 by the Smithsonian Institution
All rights reserved

Copy editor: Anne R. Gibbons
Production editor: Duke Johns
Designer: Janice Wheeler

Library of Congress Cataloging-in-Publication Data
Johnsgard, Paul A.
   Grassland grouse and their conservation / Paul A. Johnsgard.
     p. cm.
   Includes bibliographical references (p.).
   ISBN 1-58834-039-2 (alk. paper)
   1. Grouse—North America.   2. Birds, Protection of—North America.   I. Title.
QL696.G285 J629   2002   2002021012
598.6'3'097—dc21

British Library Cataloguing-in-Publication Data available

Manufactured in the United States of America
09 08 07 06 05 04 03 02   5 4 3 2 1

∞ The paper used in this publication meets the minimum requirements of the American National Standard for Information Sciences—Permanence of Paper for Printed Library Materials ANSI Z39. 48-1984.

For permission to reproduce illustrations appearing in this book, please correspond directly with the author. The Smithsonian Institution Press does not retain reproduction rights for these illustrations individually, or maintain a file of addresses for illustrations sources.

CONTENTS

Preface ix
Acknowledgments xiii

1. ONLY THE SILENCE PERSISTS: THE HEATH HEN AND NEW ENGLAND'S SCRUB OAK BARRENS 1
2. A LAST STAND IN TEXAS: THE ATTWATER'S PRAIRIE-CHICKEN AND THE GULF COAST PRAIRIES 17
3. FADING FOOTPRINTS IN THE SAND: THE LESSER PRAIRIE-CHICKEN AND THE SAND SHINNERY GRASSLANDS 29
4. A DRUMMING AT FIRST LIGHT: THE INTERIOR GREATER PRAIRIE-CHICKEN AND THE TALLGRASS PRAIRIES 52
5. DAWN DANCERS ON DUN GRASS: THE SHARP-TAILED GROUSE AND THE NORTHERN PRAIRIES AND SHRUBLANDS 81
6. DARK SHADOWS IN SILVER SAGE: ENIGMATIC GROUSE OF THE SAGEBRUSH STEPPES 104
7. CAN THE FABRIC BE MENDED AND THE PIECES PRESERVED? 123

Identification Key to the North American Prairie Grouse 135
Bibliography 137
Index 153

PREFACE

After I published *The Grouse of the World* in 1983, I thought I would never again revisit the grouse family as a subject for a book. Indeed, my attention soon turned to other avian groups, such as hummingbirds, owls, hawks, and trogons, all of which seemed to me to warrant more scrutiny and immediate conservation concern than the grouse. In my preoccupation with these subjects over the past 20 years I failed to see that a disturbing trend was developing with regard to the North American grouse, namely that their populations were almost universally declining, especially in the case of the grassland-adapted species. There are also three species of tundra-adapted ptarmigans (*Lagopus* spp.) in North America and three species of forest-adapted grouse, namely the blue grouse *(Dendragapus obscurus)*, spruce grouse *(D. canadensis)*, and ruffed grouse *(Bonasa umbellus)*, but none of these has suffered the degree of population and range declines that have occurred in the prairie grouse.

I first began to perceive this disturbing trend during my preparation of *Prairie Birds: Fragile Splendor in the Great Plains*, which was written in large measure to try to document recent habitat and population changes in some 30 species of birds that are especially associated with Great Plains grasslands. These species include the greater and lesser prairie-chickens and the sharp-tailed grouse. I gradually became aware of the increasingly critical status of the Attwater's prairie-chicken, a Gulf Coast race of the greater prairie-chicken that seems likely to follow the ill-fated prairie-chicken of the East Coast, the heath hen, into oblivion before many more years have passed.

In subsequently reviewing population data on prairie grouse populations throughout the Great Plains, I was shocked to discover they had declined sharply since my earliest attempt to summarize their national status *(Grouse and Quails of North America)* and both the greater and lesser prairie-chickens had essentially or entirely disappeared from several states during the past 30 years. These losses occurred in spite of the fact that the Endangered Species Act was passed in the early 1970s, which led ornithologists and other biologists to believe we had finally been given a cushion of protection from such catastrophic events as a species' total elimination, at least when measures could be taken to prevent it.

An explanatory note: the heath hen, the Attwater's prairie-chicken, and the greater prairie-chicken occurring in the continental interior are now all simply regarded as geographic races of one rather variable species, the greater prairie-chicken. Although current ornithological practice frowns on giving distinctive names to populations that are only geographic races, I have often had to make it clear whether I was referring to the greater prairie-chicken species collectively or only to one of these three races. Using "heath hen" for the eastern race *(Tympanuchus c. cupido)* and "Attwater's prairie-chicken" for the Texas coastal race *(T. c. attwateri)* provided an easy and well-accepted solution for those two forms, but a convenient English title for the race that historically occurred throughout the continental interior *(T. c. pinnatus)* remained problematic. Various authors have designated this form as the "prairie hen," the "western prairie-chicken," or the "northern prairie-chicken." I have settled on referring to it as the "interior greater prairie-chicken" to distinguish it from the other currently accepted races of the greater prairie-chicken whenever it has seemed necessary.

I have also followed long-held tradition, but not current American Ornithologists' Union (AOU) practice, in using English names for distinguishing the three races of sharp-tailed grouse discussed in this book, namely "Columbian sharp-tail" for *T. phasianellus columbianus*, "plains sharp-tail" for *T. p. jamesi*, and "prairie-sharp-tail" for *T. p . campestris*. Finally, the appearance of a newly recognized sage-grouse species, the Gunnison sage-grouse *(Centrocercus minimus)*, has meant that the original sage-grouse *(C. urophasianus)* needs a comparable modifying English name. I have referred to it as the "greater sage-grouse," the name recently adopted by the AOU, with its subsidiary races referred to as the northern *(C. u. urophasianus)* and western *(C. u. phaios)* greater sage-grouse.

Because essentially all the technical literature on grouse ecology and biology has been based on traditional English unit measures (miles, acres, etc.), I have used these units except for a few cases where the original measurements were given as metrics, in which case I have provided both. I have given direct (parenthetical) literature citations in captions for maps and figures, but not in the text. Where I have felt that specific statements as to sources are needed in the text, the author's or primary author's name is provided, to allow readers to locate the citation in the bibliography. Where no such source is mentioned I considered the information to be available in standard references, such as my two earlier books on the grouse *(Grouse and Quails of North America; The Grouse of the World)* or in *The Birds of North America* monographs on each of the prairie grouse species (Connelly et al.; Giesen; Schroeder and Robb; and Schroeder et al.). The drawings and maps are all my own, with documentary sources indicated as necessary.

Documenting the disappearance of a declining species from a state or region is much more difficult (and much more depressing) than observing and reporting the first appearance of an expanding species—the quantitative difference between an extremely rare species and an entirely extirpated or extinct one is very small indeed, although the biological significance might be great. The exact date of extinction has been documented for only one of the four North American bird species known to be extinct, the passenger pigeon. The last surviving passenger pigeon, after a long life in captivity, died on September 1, 1914. However, we have come close to knowing the exact date of demise for the heath hen, which likely occurred in late March or early April 1932. On April 21, 1932, after it was realized that the last surviving heath hen had disappeared and apparently died on Martha's Vineyard, Massachusetts, within the previous month or so, the *Vineyard Gazette* printed the following obituary: "Something more than death has happened, or, rather, a different kind of death. There is no survivor, there is no future, there is no life to be recreated in this form again. We are looking at the uttermost finality which can be written, glimpsing the darkness which will not know another ray of light. We are in touch with the reality of extinction."

I sat down to write this book largely as a cautionary essay, in the belief that, whereas as George Santayana noted, those who forget the past are condemned to repeat it, those who remember the past are in a better position to try to alter the future. We need to act promptly, and with a full measure of devotion, if we are to prevent the other prairie grouse from going the way of the heath hen. If we don't, future generations will point to us as just another sad example of humans conveniently forgetting the past in our usual preoccupation with the present.

ACKNOWLEDGMENTS

As with all my previous books, this one not only reflects a personal labor of love but also depended on the help and advice of countless others, whether directly given or found in the pages of a thesis, technical paper, research report, or book, often written by now-deceased authors. As Reinhold Niebuhr once wrote, "Nothing worth doing is completed in our lifetime; therefore, we are saved by hope."

In the course of my writing I relied heavily on references and information helpfully provided by various librarians, most significantly those of the University of Nebraska libraries and Barbara Voeltz of the Nebraska Game and Parks Commission library. My son Scott Johnsgard ferreted out obscure information on Texas and its wildlife and water resources, and Clait Braun offered information and helpful advice, especially about the recently discovered Gunnison sage-grouse. David Wiedenfield of the Sutton Avian Research Center in Oklahoma helped put me in touch with prairie grouse specialists through the newsletter of the Prairie Grouse Technical Council (available online at www.suttoncenter.org / PGTCNews.html).

Parts or all of the manuscript were critically read by various friends and students, including Linda Brown, Jackie Canterbury, and Josef Kren. Tom Shane located a wonderful lesser prairie-chicken lek to visit and photograph, offered valuable advice, and otherwise helped enormously with my field observations in western Kansas, as did Christian Hagen, Curran Salter, and T. J. White. Population status, hunter-harvest data, and other state and regional information on prairie grouse were provided by Roger Applegate, Rick Baetsen, James Bailey, Matthew Bain, Kelly Cartright, Elmer Finck, Ron Fowler, Joe Hartman, Tom Jervis, Douglas Johnson, Jerry Kobriger, Tony Leif, John McCarthy, Roger Peterson, Joel Satore, Steve Sherrod, Nova Silva, Brian Stotts, Daniel Svedarsky, Scott Taylor, and others.

This is not a single story; it is a collection of mostly unfinished and incomplete stories whose endings are still unknowable but are often ominous. Unlike most stories, these cannot be started over should they end badly. Except for the story of the extinct heath hen, the still-unwritten chapters may be determined by people now too young to know or care about the birds, or perhaps even by those still unborn who might come to love them and the native grasslands on which they depend. As Reinhold Niebuhr also once wrote, "Nothing we do, however virtuous, can be accomplished alone; therefore, we are saved by love."

Half the royalties from the sale of this book will be assigned to the National Audubon Society.

1

ONLY THE SILENCE PERSISTS
The Heath Hen and New England's Scrub Oak Barrens

The morning of March 11, 1932, dawned cold and probably foggy on Martha's Vineyard, a largely sand-covered island of about 50 square miles just off the southeastern coast of Massachusetts. The morning was much like thousands of other March mornings. And like many March mornings of the previous few years, it was greeted by the soft, sad sounds of a lone male heath hen proclaiming his ownership of a tiny territory in a low, grassy meadow near Edgartown. He was beginning what was apparently his final performance, in a long-running act that had begun in obscurity and was about to end in silence.

Whether the heath hen was native to Martha's Vineyard or had been transplanted there from the Massachusetts mainland (about five miles distant) by early colonists is not known. Suitable brushy habitat, with many scrub oaks and fruiting heath shrubs such as blueberries and huckleberries growing in the sand barrens, had certainly been present since at least the early 1600s, when the island was first well described by European explorers. In 1916 as many as 2,000 heath hens lived on this small, often windswept island—the only surviving birds anywhere. This amazing population density, representing about 40 birds per square mile, would never again occur. The birds had been gone from the New England mainland since about 1840, despite having been protected from hunting on the Massachusetts mainland in varying degrees since 1831. Full and effective protection on Martha's Vineyard was not implemented until 1907. Then a privately funded preserve was established, which increased in size to more than 2,000 acres by 1911 (see Map 2, p. 11). The preserve was located immediately east of Martha's Vineyard State Forest, in scrub oaks and pines (scrub coppice forest), with denser woods to the north and with open grasslands and agricultural lands to the south.

The earliest known islandwide survey of heath hens in 1890 indicated that 120 to 200 birds were then present. A fire on the breeding grounds in 1894 decimated the population. A similar fire occurred in 1906, and in the following spring only 21 birds were found on the entire island. Annual counts were initiated in 1906, and by 1908 between 45 and 60 birds were documented. These annual counts suggested that a cyclic population pattern existed on the Vineyard, marked by a low point of fewer than 60 birds in 1908, followed by an amazing increase only a few years later to 2,000 birds. Probably this remarkable change in status resulted from the establishment of a sanctuary in the middle of the bird's breeding range, as well as the effective elimination of foxes and raccoons from the island. The sanctuary eventually encompassed some 1,600 acres and was supplemented by the hiring of permanent wardens to help protect the birds.

Another disastrous spring fire, at the peak of the birds' breeding season, swept over the island in May 1916, leaving perhaps as few as 150 survivors. In 1920 an epidemic of blackhead, a disease of poultry that had been introduced on the island by domestic turkeys, seriously weakened the heath hen flock, which by then had increased to about 600 birds. In 1923 only about 50 birds could be found, and in 1924 even fewer were present. Few if any broods were documented after 1925. By the spring of 1929 only a single male could be found on the entire island.

This male faithfully began his lone territorial rituals in March 1929 and appeared again in the same field to repeat them regularly but entirely alone during the spring of 1930 and 1931. By 1931, when he was caught and banded, he was believed to be perhaps seven years old, a notable age for a wild grouse of any species. More remarkably, he reappeared yet again on the display ground in the early spring of 1932. He was now perhaps eight years old, a centenarian by human standards. He probably welcomed that dawn of March 11, 1932, with the same mournful sounds that he alone had uttered during the three previous springs, again without receiving a single challenging reply. It was apparently the last time any person, or indeed so far as is known any living being, would ever hear the territorial call of a male heath hen.

We often think the early colonists of Massachusetts must have subsisted on a regular diet of wild turkeys, for that is how the paintings and stories of our youth depicted their first Thanksgiving celebration. Yet it was probably the heath hen that most often appeared during their Sunday, and perhaps weekday, meals. The birds were so abundant in eastern Massachusetts during the early 1600s that they could be purchased at markets for four pence each, and it was said that a hunter could readily kill half a dozen birds in a single morning. They were still so abundant around Boston in the late 1700s that servants and common laborers arranged a formal agreement with their masters and employers that they would not be required to eat heath hen meat "more than a few times in the week." This state of affairs would not last; for as the human population increased and yearlong hunting remained uncontrolled, the seemingly limitless numbers of heath hens began to decline. Additionally, the "barrens" haunts of the birds were among the first to be colonized by settlers, for these could quite easily be cleared of unwanted woody vegetation, and those that weren't soon began to grow into larger-stature woodlands as they were protected by the settlers from frequent fires.

The downward population trend of heath hens brought about by this combination of overhunting and progressive habitat loss apparently was occurring rather rapidly by the early 1800s. In the 1830s heath hens were still to be found on Long Island, in the coastal plains of New Jersey, in the Connecticut River Valley, and in mainland Massachusetts, as well as on Martha's Vineyard. The last record of the birds' being shot in western Massachusetts was in 1830, but within 10 years they were apparently gone from the entire Massachusetts mainland, as well as from all of Connecticut. Heath hens became increasingly rare on Long Island in the late 1700s and began to receive legal if inadequate seasonal protection from hunting during spring and summer in 1791. During the first two decades of the nineteenth century the price of heath hens there gradually increased from $1 to $5 each, a high price indeed for any game bird at the time. By 1844 the heath hen was thought to be virtually gone from Long Island, and it was certainly gone from the rest of New York by then. The last specific New York record for the heath hen was apparently about a decade earlier, around 1835. By about 1850 heath hens were also disappearing from the scrub oak uplands in the Pocono Plateau of northeastern Pennsylvania. By the 1830s the birds were selling in the Philadelphia markets for $5 to $10 a pair, according to J. J. Audubon. Specimens were collected in Lancaster County, Pennsylvania, sometime before 1850 and in Monroe County about 1860. The last sight records for that state were from Monroe and Northampton counties, around 1869.

On the sandy pine-oak plains of southern New Jersey (Monmouth and Burlington counties), the birds may have remained fairly common up until about 1850. The last known specimen from New Jersey is believed to have been obtained some time before 1892, decades after the birds had disappeared from everywhere else on the East Coast except for Martha's Vineyard. Meantime, in Chicago, hundreds of thousands of the closely related greater prairie-chicken were still being sold annually by market hunters for about $3.50 a dozen. Southward, beyond the pine barrens of New Jersey, the picture is unclear, but a heath hen specimen was collected near Washington, D.C., in 1846, and one was reported from Prince Georges County, Maryland, in 1860. As for Virginia and

North Carolina, there is no clear evidence that the heath hen ever extended that far south, in spite of some early assertions that it may have. John Aldrich provided what is perhaps the best speculative range map of the presumed distribution of the heath hen, which included in its range extreme northeastern Virginia opposite the Potomac River. Beliefs that its range extended still farther south may have been the result of the rather indefinite geographic limits of both "Virginia" and "Carolina" that were implicit in early colonial writings, as well as the similarly informal and often interchangeable use of terms such as "grouse," "heath-hen," and even "pheasant" for various game birds. Certainly acorn-bearing scrub oaks extend as far south as Virginia and North Carolina on poorer sandy and acidic soils, but the associated fruit-bearing shrubs such as blueberries and huckleberries that were used by the heath hen for important late summer and fall foods become less frequent in the southern Appalachians.

HISTORIC RANGE AND LANDSCAPE ECOLOGY

Unfortunately, the last heath hen died before ecology had really been born, and we have little direct evidence as to the birds' actual habitat needs. This eastern bird, a minor plumage variant of the interior and Attwater's greater prairie-chickens, actually differed in many ecological respects from its much more widespread relatives of the western Great Plains and Gulf Coast prairies. The vast expanses of grassland that these latter populations now require have not occurred in the eastern states during the past several centuries; thus biologists are prone to wonder how the scrub- and heath-adapted heath hen could ever have evolved from its open-country and grassland-dependent ancestral relatives. Yet during the late Pleistocene period, namely toward the end of the last glaciation some 18,000 to 12,000 years ago, large areas of nearly treeless parkland occurred just south of the glaciers that were by then generally stable or retreating. These regions consisted of a vast tundra- or parkland-like community, with scattered spruces present in a grassland matrix that extended across much of what would later become the northern and northeastern forests of the United States.

All grouse species are well adapted to life in cold, even arctic, environments. All have feather-covered nostrils that, scarflike, probably help keep frigid air from entering the respiratory system. By comparison, other gallinaceous birds, such as the native American quails, have bare nostrils. Additionally, all grouse have their lower legs (tarsi) covered with feathers, usually to the base of the toes, and in the arctic-adapted ptarmigans all the way to the tips of the toes. The most cold-adapted grouse species have longer tarsal feathers, whereas species (and races) adapted to warmer climates have this feathering variously reduced. Finally, grouse develop a fringe of comblike scales on both sides of their

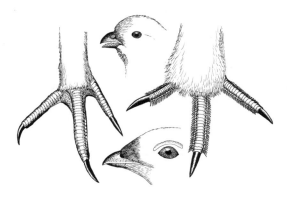

Figure 1. Legs and toes of greater prairie-chicken in summer *(left)* and plains sharp-tailed grouse in winter *(right)* to illustrate leg feathering differences and seasonal toe pectinations. Also shown are the heads of a representative grouse *(below)* and a New World quail *(above)*, to illustrate differences in nostril feathering.

toes each autumn; these "pectinations," which seemingly vary in size not only by species but also by geographic race, evidently serve in the manner of snowshoes by helping to spread the weight of the birds as they walk over soft snow. All these attributes are illustrated in Figure 1.

As an example of Ice Age range displacements of grouse and other birds, in Pennsylvania a fossil assemblage from 11,000 years ago included a skeleton of the sharp-tailed grouse, a bird now found only several hundred miles to the west. A Pleistocene site of similar age from Virginia also included the fossilized remains of a sharp-tailed grouse, as well as such present-day western grassland birds as the black-billed magpie and brown-headed cowbird. There is also a fossil species of probable prairie-chicken, *Tympanuchus lulli,* from Monmouth County, New Jersey, the heart of the heath hen's historical range. That Pleistocene-age bird might have been somewhat larger than the present-day forms of greater prairie-chickens, as are many other birds and mammals that date from that same cold epoch.

Summarizing the evidence for the presence of a widespread postglacial grassland community in eastern North America, Robert Askins noted that at least three birds now associated with grasslands much farther west developed endemic eastern races, probably during late glacial history. They include the heath hen, the eastern race *(susurrans)* of Henslow's sparrow, and the Ipswich race *(princeps)* of the Savannah sparrow. The eastern grasshopper sparrow's present range is especially similar to the historical race of the heath hen, extending from southern New England south to North Carolina. Quite possibly the Florida race *(floridana)* of the burrowing owl may also have become isolated from its western relatives at about this time or somewhat later, during a prolonged period of postglacial warming and drying that occurred less than 6,000 years ago. Robert Askins also mentioned three eastern vascular plant species (golden aster, bushy rock-rose, and sandplain agalinis) and one additional subspecies of grassland forb (blazing star) having their closest geographic affinities with the present-day Great Plains grasslands.

After the final retreat of the glaciers, starting about 10,000 years ago, the eastern parts of North America began progressively to assume their present-day shape. Probably most of the true eastern grasslands and parklands gradually disappeared and were replaced by forests of various types, with some exceptions. One of these was the retention of coastal grasslands on nonstabilized sands close to the ocean itself, a habitat type exploited by the Ipswich race of the Savannah sparrow. Additionally, more widespread, open communities formed over relatively infertile, often sandy, substrates that did not allow a forest to develop. Instead they became colonized by scrub oaks, similarly scrubby pines, and various shrub species including many in the ericad, or heather, family. These scrubby woodlands or forest openings, variously called oak, pine, blueberry, or huckleberry barrens, occurred from eastern Maine southward into the central Appalachians. They have been especially well studied in eastern Massachusetts, central Long Island, and southern New Jersey, all important historical habitats for heath hens. Such scrub-dominated barrens often developed over well-drained and rather coarse, sandy, and infertile soils that are derived from glacial outwash, but also sometimes occurred on unglaciated substrates derived from sandstone.

Over most of the heath hen's presumed historical range its known limits occurred within the region of glacial influence. In New Jersey the southern edges of the Kansas-Illinoian ice sheets are located in the northern part of the state, just short of the extensive New Jersey pine barrens. These barrens were subsequently formed by the vast quantities of sands and gravels deposited to the south, as the ice sheets melted and retreated. This process was well summarized by B. Robichaud and M. Buell *(Vegetation of New Jersey: A Study of Landscape Diversity)*. Most of this large region consists of New Jersey's outer coastal plain, which comprises about 3,500 square miles, from Monmouth south to Cumberland and Cape May counties. This entire region is now mostly less than 100 feet above sea level, and its soil substrate is little more than sand, clay, and gravel. Several different plant communities occur within this region, which is collectively called the pine barrens. They are well described in *Pine Barrens: Ecosystem and Landscape* (edited by Richard Forman). Each of these major communities largely depends on the relative frequency of fire. Pine-dominated forests of moderate to substantial height tend to grow where the fires occur at intervals of no less than about 20 years. In this forest community, pitch pine is usually dominant, but small-stature oaks (mainly bear oak) are common, as are blueberry, huckleberry, and other woody heaths. In areas where the fire frequency occurs every 5 years or so a chaparral-like dwarf pine community develops, with pitch pines, bear oak, and blackjack oak variably present. Here the shrub component is especially rich in huckleberries and blueberries. Both of these, and other important food sources, are members of the heath family, whose species are notable for being able to flourish on the poorest and most sterile of soils because of nitrogen-fixing fungi that are closely associated with heath root systems. These stunted-forest "pine plains" now occupy about 15,000 acres of New Jersey, and very likely represented the heart of New Jersey's heath hen range, the abundant acorns and heath-based berries providing the bird's prime food and winter needs, and the thick shrubbery providing nest, escape, and winter cover. A similar pattern of vegetation has been described by Linda Olsvig and others in "Vegetational Gradients in the Pine Plains and Barrens of Long Island, New York." Although this island was once largely covered by forests, a prairielike area dominated by grasses (called Hempstead Plain on the west) graded into pine forests toward the east, through an intermediate community type known as the oak-brush plains and somewhat taller and denser pine barrens. In both these community types, bear oak comprises an important shrub layer, and heaths such as blueberry, bearberry, and huckleberry are quite common.

In southeastern Massachusetts and also on Martha's Vineyard, a closely comparable community called the pitch pine–scrub oak barrens once occurred widely, and remnants still exist on undeveloped sites. It also occurs as a boreal variant as far north as extreme southeastern Maine and adjacent New Hampshire, and in eastern New York, scattered parts of Pennsylvania, central Massachusetts, and adjacent Connecticut. Coastally situated variants such as those in Rhode Island, Long Island, and southern New Jersey closely approximate those of southeastern Massachusetts and Martha's Vineyard. In all these regions the soil substrate consists of little more than relatively infertile sand, and the primary tree is usually pitch pine of varying heights. Scrubby-stature bear oak is also present and generally more abundant near the coast, whereas chinquapin oak becomes more frequent inland. Huckleberries are usually abundant, supplemented by lowbush blueberries and bearberries. A few perennial grasses typical of western prairies, such as little bluestem, may also occur in open sites. Frequent fires helped maintain the typical low woody stature of this community type and probably made the habitat more readily accessible for food and escape cover by heath hens. Since the 1930s the elimination of these fires has allowed these scrub communities to grow into full-size forests, with nesting great horned owls, red-bellied woodpeckers, and other birds typical of the taller forests of New England.

A final important type of eastern grassland was that generated from forests and retained as open lands for varied periods by human activities, most often by forest clearing or by setting fires for clearing brush or driving game. Although one might imagine such practices to be primarily those associated with European colonists, the Native Americans clearly engaged in similar practices even earlier. Some naturally occurring and repetitive fires may also have helped prevent the eastern forests from becoming the dark, deep wilderness we might like to imagine as typical of precolonial eastern North America. Stephen Pyne's *Fire in America* has documented the fact that setting fire to the environment is as much a traditional American practice as making apple pie. Gordon Day's review of the role of Native Americans in influencing vegetation patterns in the Northeast through their use of fire is likewise of great interest. From at least the early 1600s Native Americans were known to set fires frequently, sometimes during both spring and fall, evidently in large part to clear out undergrowth along game trails and thus facilitate their seasonal hunting for deer and other game.

One well-known example of a natural or possibly fire-maintained secondary type of grassland was Hempstead Plains, in Nassau County, western Long Island. It has been described as a treeless, grassy plain from times as early as the 1600s, and until the 1930s this area of western Long Island still botanically resembled the tallgrass bluestem prairies of western Iowa and eastern Nebraska. Breeding bird species typical of these more western prairies, such as upland sandpipers, vesper sparrows, and grasshopper sparrows, were present as recently as the 1930s. Historically, the heath hen was also present here, according to Robert Askins. This area of some 50,000 acres (78 sq mi) may well have been one of the largest natural grasslands occurring along the entire East Coast, but other smaller and apparently natural or perhaps fire-maintained grasslands also occurred on eastern Long Island, in the mountains of northeastern Pennsylvania, and in Connecticut. Farther inland, from western New York through the Great Plains states, species of oaks formed oak-dominated barrens or oak savannas. These communities are often located near the southern terminus of glacial advances, where terminal moraines and sandy or gravelly outwash plains are common. There is no evidence that the heath hen's range ever extended farther west than central Pennsylvania.

It is possible to produce some rather hypothetical maps of the historical distribution of the heath hen, based on the relatively few actual specimens or reliable literature-based localities known for it, the known distribution of pine-oak barrens in the Northeast, and the distribution of important food plants of the heath hen. Tallying the documented New England county distributions of nine known major food plants as reported for the heath hen, mainly the acorns of bear oak and the fruits of a variety of mostly ericaceous shrubs, especially *Vaccinium* species, as well as some other important berry-producing food plants such as bearberry *(Arctostaphylos)* and partridge berry *(Mitchellia)* produces a probability-of-occurrence map (Map 1). Comparing this map with the known

Map 1. Historical distribution of the heath hen (solid line). Dashed line shows probable extension of the heath hen's distribution into southern Maine. Inked-in areas indicate locations of major historical scrub barrens or coastal prairie habitats within heath hen's known and probable distributions; stippling shows pine-oak barrens outside heath hen's known historical range. The insets show a leaf and acorn of bear oak *(above),* and leaves and berries of blueberry *(below left)* and partridge berry *(below right),* all important fall and winter foods.

approximate distributions of pine-oak barrens in the Northeast and with the available heath hen specimen records suggests a similar historical distribution (Map 2).

The heath hen evidently reached its westernmost limits in Pennsylvania. It was certainly common on the Pocono Plateau and Broad Mountain in the northeast, these areas lying within the southern limits of the Wisconsinian glaciation. It also extended beyond the glaciers' limits, south to York, Lancaster, and Chester counties, and west apparently at least locally to Clinton and Union counties. The birds were reportedly limited to the pine and scrub oak barrens of eastern and central Pennsylvania, and may have been prevented from further western expansion there by the more heavily wooded and colder Appalachian highlands.

The maps included here would suggest that the heath hen's northern range may once have reached southwestern Maine, as has been suggested by such early writers as Alfred Gross and by more recent ones such as Roger Applegate. Furthermore, as Applegate noted, the surviving pine-oak barrens of southern Maine still support remnant populations of essentially the same breeding birds (grasshopper sparrow, vesper sparrow, and upland sandpiper) as still occur in the barrens farther south, and they may once have also supported the heath hen.

SOCIAL AND SEXUAL BEHAVIOR

By the early 1900s, when the entire population of heath hens had been reduced to those on Martha's Vineyard, the island was visited by many biologists for a chance to view these rare birds. A small wooden observation blind was constructed about 1920, in a grassy farm field just east of West Tisbury. This field was the birds' favored if not only communal display site during their final decade or so of existence. At about this time Alfred O. Gross began his exhaustive studies on the birds. A Harvard graduate, he had only recently been promoted to the rank of full professor at Bowdoin College, a small liberal arts college in Brunswick, Maine. He first visited Martha's Vineyard in 1923, and continued to study heath hens there until 1932, when the last bird disappeared. The following account is based mainly on his monographic writings, originally published by the Boston Society of Natural History. Those accounts were later supplemented by a summary that he provided for A. C. Bent's classic *Life Histories of North American Gallinaceous Birds,* published in the same year the last heath hen died.

The heath hens on Martha's Vineyard typically began their spring display season in late February or early March. During eight years when records were kept, displays began as early as February 12 and as late as March 9. Maximum

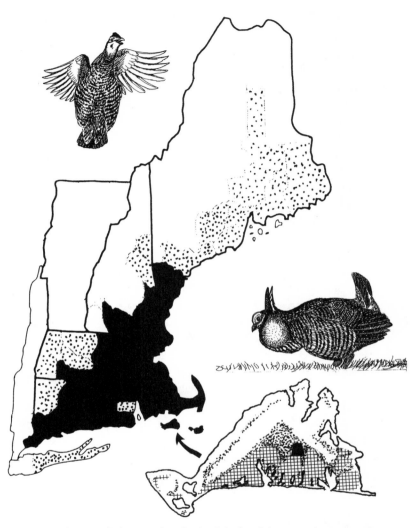

Map 2. Distribution of 9 known plant foods of the heath hen (mainly species of Ericaceae, plus bear oak) in northern New England. Counties with records of all 9 species are inked; those with 7–8 are stippled (based on Magee and Ahles 1999). The inset map of Martha's Vineyard shows the locations of native grasslands (cross-hatched) and scrub coppice thickets (stippled); unmarked areas are nearly all taller woodlands and forests (after Ogden 1961). The historical heath hen preserve is also shown (inked).

frequencies of displays occurred between the first of April and the second week of May, and probably most matings also occurred then, although Gross never was lucky enough to observe one. During four years, the last seasonal displays occurred between June 2 and June 20. These are surprisingly late dates compared to observations of greater prairie-chickens at similar latitudes in the midwestern states (central Nebraska), where displays typically cease by early May. It is possible that these very late dates for Martha's Vineyard reflect the rarity or absence of females in most years, forcing the males to remain on their display grounds much longer than normal. The few available nest records are for June and July, which are also rather late dates relative to Nebraska.

Based on Professor Gross's and other earlier accounts, no clear differences between the heath hen's social behavior and those of the other types of greater prairie-chickens can be detected, and little purpose is served in providing extended descriptions that do not directly answer the question of whether such differences may actually have occurred. The basic strategy of male heath hens, as with all the prairie grouse, was to try to establish and advertise a mating territory that would be successful in attracting as many females as possible for fertilization. This means the bird must locate itself on a site with broad visibility, bringing attention to the male not only from females but also from predators. The "drumming fields" of the heath hen fulfilled this role; it was typically an open, flat meadow, covered by low grasses and forbs.

The hours around dawn are especially effective for attracting females but avoiding predators, since by then most owls have ceased hunting but hawks have not yet become very active, providing a temporal window of opportunity. Effective communication signals available during these hours include acoustic ones such as vocalizations as well as mechanical sounds such as foot-stamping and feather rustling. Both these sound types work fairly well at distances of up to about 25 yards, but for long-distance communication a loud, very low-frequency, and prolonged sound works best, similar to that produced by a foghorn or tugboat. The male heath hen's primary call filled this need perfectly. Alfred Gross noted that he could hear the "booming" calls from a distance of at least two miles from the display grounds under favorable conditions. He reported that one sequence of 42 such calls was uttered by a single male within a period of about 10 minutes, interspersed with 10 cackling notes. These cackles were used in close-range male-to-male aggressive encounters, whereas the booming calls seem to have had the dual purpose of attracting females and perhaps repelling other males.

Professor Gross also provided some detailed information on the anatomical mechanism of the male's remarkable primary territorial advertisement call. In preparation for the call, the tail was cocked; the outer (primary) flight feathers of the wings were slightly lowered; the neck was stretched forward and upward; and the earlike pinnae were variably erected, often directly in diagonal line with the neck. The bare patches of yellow skin above the eyes were enlarged through vascular engorgement to form comblike "eyebrows." The lateral orange-colored "air sacs" were then inflated, so that they resembled two hemispheres (Figure 2). A low-pitched, three-syllable call was also produced.

Curiously, the beak was held tightly closed during almost this entire performance. Additionally, the tongue was evidently lifted to the palate so that it closed the entrance to the internal nostrils (nares), preventing both air and sound from escaping readily through the external nostrils. As air was expelled from the lungs, it caused the vibratory membranes located at the junction of the windpipe and bronchi (part of a unique avian structure called the syrinx) to vibrate. These vibrations of varied frequency and amplitude generate all the vocalizations of birds, including relatively simple calls and the generally more acoustically complex songs. The air, unable to escape easily through either the

Figure 2. Booming postures of male heath hen (*above;* after photo by Alfred Gross) and Attwater's prairie-chicken (*below;* after photo by John Tveten).

closed beak or the nostrils, inflates the anterior end of the esophagus. Thereby the surrounding neck skin is exposed, and the rounded air sacs are progressively enlarged and more conspicuous.

Furthermore, the large air volume trapped in the esophagus acts as a low-frequency resonator and amplifier for the sounds generated by the syrinx. As a result, a rather muffled, almost kettledrum-like sound is generated that, although of limited volume, has remarkable carrying power. Gross described the resulting call as sounding like "whoo-dooh-doooh," "whhoo-ooo'dle-dooooooh," or other variants. Similar interpretations exist, of which "Old-Mul-dooon" has the double advantage of being fairly close phonetically and also easily remembered. According to Professor Gross, the entire performance required from 1½ to 3¼ seconds for the male to complete. This is a surprisingly variable time period, given the quite consistent time intervals typical of the other races of greater prairie-chickens.

Like calls, variations in general body postures and associated feather-outlines can be effective signals under low-light conditions, so tail-cocking and the erection of pinnae may be useful at such times. The exposure of contrasting white feathering (such as exhibiting the under tail coverts by tail-cocking) also works rather well under conditions of reduced light. Exposing brilliant color patterns such as iridescent feathers would not be so effective as using this type of long-distance call, as these are rather hard to see in dawn light. However, early light, especially that occurring around the time of sunrise, is rich in long wavelength hues that are centered around red, so any important visual signals used at that time of day should most effectively use this part of the color spectrum. The orange bare patches of inflatable neck skin (air sacs) and the bright yellow eye-combs are thus ecologically adapted for effective exposure during this time of day. When not in full display, both these structures become virtually invisible, reducing the birds to a confusing zebralike striped pattern of buff and brown, perfectly adapted to concealment in a dead-grass environment.

The last and certainly a remarkable feature of social behavior typical of the heath hen, indeed one that is present in all the prairie grouse, is that the reproductively active males display simultaneously in competitive arenas. In this strange strategy, all the adult males of a local population interact competitively in a single location at the same time, in order to establish which of all the males present is the most dominant and thereby determine which will have first access to the available females. The remaining males also sort themselves hierarchically by relative dominance, thus establishing a positional relationship of their territories with respect to each other and to the most dominant male. The result is a generally concentric pattern of males that is a direct reflection of their individual social status. The central and most dominant male, the "master cock," is only infrequently challenged directly as to his position or first-rights access to females, at least after his rank has been firmly established. The incidence of such challenges and actual fights increases proportionally among males situated at increasing distances from the arena's center. In most arena-forming grouse it may require several years for a male to "work his way up" to attaining the status of master cock. Because of the high mortality rates of grouse in general, and perhaps also because of the possible stresses associated with maintaining this status, only a few males ever manage to hold this exalted rank for more than a year or so before dying or being displaced.

This gladiator-like type of reproductive strategy is quite rare in birds; less than 2 percent, or fewer than 200 out of nearly 10,000 species, of the world's birds engage in it. The sites where such encounters occur are traditional, at least in the sense that the same ones are used year after year, decade after decade, as each new generation of males successively learns specific locations from the previous one. Each male automatically passes this information on to immature males of the next generation, in a manner curiously similar to the human transmission of information as to locations of special religious or cultural significance for the local population. The sites used by heath hens were generally called "drumming grounds," "booming grounds," or "tooting grounds." Females probably learn their locations by the collective sounds generated by all the participating males. However, they only visit them when they are individually ready to mate, and then only long enough to become fertilized.

REPRODUCTIVE BIOLOGY

Few observations on the later phases of reproduction were ever obtained for heath hens. In spite of repeated efforts, Professor Gross never managed to find a nest. One clutch of six well-incubated eggs was collected in Martha's Vineyard on July 24, 1885. Another clutch of nine eggs was found June 2, 1906, and one with an incomplete clutch of four (later increased to eight) eggs was located on June 5, 1912. This last nest was found in a mass of ferns, with nearby scrub oak cover about 2–3 feet high. On this occasion the incubation period was determined as 24 days, essentially the same as for the other races of greater prairie-chicken. The year 1924 marks the last spring that any young birds were seen on Martha's Vineyard.

There is also a paucity of information on their nesting and brooding success. In 1913, 9 broods had an average of 4 chicks per brood; 7 broods in 1918 had an average of 5 chicks; and in 1922, 5 broods were counted, ranging from 4 to 8 young. Other reported brood-size counts from various years were of 2 (1 brood), 5 (2 broods), 6 (3 broods), 9 (1 brood), and 10 chicks (1 brood). The average of 22 of these broods is 5.4 chicks, so it would seem that 5 or 6 chicks represented a fairly typical brood size for heath hens. These numbers are not noticeably different from brood sizes that have been reported for midwestern greater prairie-chickens and offer no special insight into the bird's disappearance. Increasing sterility resulting from forced inbreeding has been suggested as a possible cause for the bird's disappearance, and this is certainly a possible if unproven contributing factor. Supporting evidence for this idea has recently been obtained for small remnant populations of Illinois greater prairie-chickens.

American crows, feral cats, and rats were all implicated by Professor Gross as likely or known predators of eggs and chicks on Martha's Vineyard during their final years there. Among other potential predators, northern harriers and northern goshawks were thought by him to perhaps have significant effects on adult heath hens. Of these raptors, the northern goshawk is often a grouse specialist wherever it is common, but the northern harrier is normally only a small rodent specialist. Other hawks and owls occurring on the island were believed by Professor Gross to perhaps represent rather insignificant population influences. Fires during the nesting season are known to have had a devastating effect on both adults and young, and poultry diseases such as blackhead that were introduced into the heath hen's population by domestic fowl nearly spelled the end of the heath hen's existence on Martha's Vineyard in 1920.

In spite of nearly a century of conservation efforts by the state of Massachusetts, the long-running performance of the heath hen evidently came to an end sometime in March 1932. The United States was then falling into an ever-more serious economic depression, and few people took time to mourn the passing of yet another bird, less than two decades after the passenger pigeon *(Ectopistes migratorius)* and the Carolina parakeet *(Conuropsis carolinensis)* had disappeared. With the benefit of some 70 years of hindsight we would like to think we could have prevented its demise, if only we had better controlled the fires, the diseases, and all the other variables. Yet the story of the heath hen reminds us that, in the end, we are sometimes unable to change the flow of history in spite of our technology and our best efforts, and can only dutifully if sadly record it.

It is a melancholy thought that, after its compatriots had disappeared, the last surviving male heath hen in North America faithfully returned each spring to its traditional mating ground on Martha's Vineyard, Massachusetts, where he displayed alone to an unseeing and unhearing world. Finally, in the spring of 1932 he too disappeared. With him died the unique genes that reflected the sum total of the species' history, from Pleistocene times or earlier through uncounted generations of successful survival to the very last, when inbreeding, habitat disruption, fire and disease inexorably tipped the balance one final time. No one knows exactly when or how that last survivor died, and no bells tolled to mourn his passing. Indeed, only by the absence of his dirge-like booming on a March morning in 1932 was the heath hen's extinction finally established, and the bird that had been as much a part of our New England history as the Pilgrims was irrevocably lost. (Paul Johnsgard, *The Grouse of the World*)

2

A LAST STAND IN TEXAS
The Attwater's Prairie-Chicken and the Gulf Coast Prairies

There was a time when the Gulf Coast prairies must have flourished in an uninterrupted and broad green ribbon extending along the coastal plain from the Mexican boundary to Louisiana and eastward beyond the mouth of the Mississippi River. The native tall grasses of bluestems, Indian grass, and switchgrass occurring there were largely the same as those of the interior tallgrass prairies to the north, although immediately adjacent to the coast these graded into rank moisture-loving species such as cordgrass, as well as rushes and sedges that are more typical of brackish and saltwater estuaries. These latter wetlands had their own characteristic breeding rails, gallinules, and ducks, as well as distinctive endemic races of the marsh wren *(thyrophilus)*, seaside sparrow *(sennetti)*, and red-winged blackbird *(littoralis)*. In the somewhat drier and more inland bluestem-dominated grasslands, similarly endemic races of the eastern meadowlark *(hoopesi)*, horned lark *(giraudi)*, and the Attwater's race of the greater prairie-chicken also evolved along these coastal plains. These last three forms are generally thought to have become separated from their more northern related populations of the Great Plains and central lowlands during the final stages of the Pleistocene, when bird populations were still displaced varying distances to the south.

With regard to this proposed evolutionary scenario, Darrell Ellsworth and others analyzed the DNA from mitochondrial genes in the prairie grouse, including samples from the Attwater's and the more northerly populations of greater prairie-chickens, lesser prairie-chickens, and sharp-tailed grouse, testing the hypothesis that these populations may have first become isolated during the last (Wisconsinian) glacial period. Surprisingly, the data they obtained were not consistent with a presumed Pleistocene isolation of sufficient duration to achieve species-level distinction, even for the sharp-tailed grouse. The researchers suggested that postglacial population fragmentation may account for the high level of genetic similarity among them, with sexual selection alone perhaps being responsible for the fairly substantial morphological and behavioral differences evident in some of these geographically isolated forms.

The present-day differences between the Attwater's race of the greater prairie-chicken and the race occurring immediately to the north are not great. Specimen records from historical times in Texas place the two forms less than 100 miles apart, with no clear ecological barrier between them. Their plumage differences, however, are at least as great as those separating the heath hen and interior prairie-chicken, these latter two being separated by a much greater distance and much more obvious ecological barriers.

The Attwater's prairie-chicken is distinguishable from both the heath hen and the interior prairie-chicken populations by having the lower legs (tarsi) unfeathered partway up the front surface (see Figure 2). The tarsi also are unfeathered for a greater length along the rear, except during winter, when feathering extends to the base of the toes, at least in the front of the foot. Given the mild winters of the Gulf Coast, a gradual loss of tarsal feathering might be expected. Beyond the leg feathering differences, the birds average slightly smaller in most measurements than the more northern prairie grouse populations, representing another difference that can be easily related to their mild climatic conditions. They are also slightly darker and more chestnut-toned above than are northern prairie-chickens, another adaptation that is altogether appropriate to life in their more humid environment.

Additionally, the erectile pinnae of the adult males (and to a lesser degree of the females) are somewhat shorter and narrower in the Attwater's prairie-chicken than in the northern prairie-chicken (Figure 3). A logical explanation for this difference, other than sheer chance, is harder to provide. Variations on male pinnae length might reflect the intensity or effectiveness of sexual selection during courtship, namely the degree of competition among males in attracting females and the relative importance of pinnae length in attracting them. If such is the case, regional differences in this feature might well be expected. Conversely, the males of lesser prairie-chickens have relatively the longest pinnae of all the prairie grouse, another seemingly inexplicable trait.

The Attwater's prairie-chicken was named in honor of H. P. Attwater, a Texas conservationist who spent many years trying to attain protection for the birds in the early 1900s. Such complete protection only occurred in 1937, some six years after Attwater's death, by which time the birds were already in precipitous decline. The year 1937 also marks the time that Valgene W. Lehmann began his long-term study of the Attwater's prairie-chicken. Val Lehmann was then a young biologist who was collaborating with Texas A&M College, various state conservation agencies, and the U.S. Fish and Wildlife Service. He was assigned the complex task of studying all aspects of the Attwater's prairie-chicken and providing guidelines for its management and conservation. By the late 1930s the Attwater's prairie-chicken seemed to be on the way to extirpation, as had occurred with the northern prairie-chickens in central and northern Texas. These latter birds had numbered perhaps half a million strong in the 1850s but had been subject to such uncontrolled slaughter that by the turn of the century they were nearly gone from the state. The last small Texas flock was seen in 1920.

There are no estimates of the original or maximum size of the Attwater's population. In Texas alone, its total historical range of roughly 6 million acres (9,400 sq mi) was nearly as large as the heath hen's, at least in linear extent. Within this large area, the lower prairies that were dominated by tall and thick stands of cordgrass were probably seldom used, and then only for winter cover. Nevertheless, about 6 million acres of bluestem prairie were at the heart of its year-round range. We might easily imagine, assuming a moderate grouse density of only 1 grouse per 50 acres, or about 13 birds per square mile, that its to-

Figure 3. Attwater's prairie-chicken, adult male (after U.S. Fish and Wildlife Service photo).

tal population could once have numbered 100,000–125,000 birds. At its historical peak, the Attwater's range probably included parts or all of 27 Texas counties and extended east to include several western Louisiana parishes (Map 3). However, by the late 1930s the remaining range had been reduced to a handful of isolated populations. These were mostly located within 7 Texas counties, extending from Refugio on the south to Jefferson on the north. The last Louisiana birds had disappeared by 1919.

Val Lehmann engaged in a careful survey of all the remaining Attwater's prairie-chicken habitat left in Texas. As of the fall of 1937, he estimated that 457,000 acres of grouse habitat still existed, or only 7 percent of the original 6 million acres. The total prairie-chicken population was then estimated at about 8,700 birds, a reduction of more than 99 percent of the number he imagined may once have existed (evidently nearly a million birds). Almost half the total known population (4,200 birds) was then limited to two private ranchlands in Refugio and Aransas counties, where the estimated population density was about 1 bird per 10 acres. Statewide, the average grouse density was only a bird per 52.4 acres in the areas Lehmann surveyed.

In that same year, the state of Texas declared a closed season on prairie-chickens for five years. In several counties essentially unregulated hunting had been cited by Lehmann as the factor largely responsible for the birds' decline, and few game wardens were present at that time to help insure that the existing regulations were being followed. From 1925 through 1937 a season limit of 10 prairie-chickens had been allowed, but the early season (in September) resulted in high losses of newly fledged young, as well as still-molting adults.

Overgrazing, pasture-burning, and mowing, conversion of native prairies to cropland, oil development, and losses of surface water through drainage were believed to be additional significant factors brought about by human activities that affected prairie-chicken populations. Nearly 3 million acres of land in the range of the prairie-chicken had been converted to cropland by 1936. Chief among these crops was rice, which offered few potential benefits for prairie-chickens in terms of either new food resources or cover for nesting or escape. One area, which had supported a population of about 10,000 prairie-chickens before rice culture began in 1924, had only 150 by 1937.

Natural factors, including droughts, severe weather during the breeding season, the encroachment of brush onto prairie habitats, and predators, were also considered as important mortality factors in some locations at some times. At the time of Lehmann's study, skunks, opossums, raccoons, red wolves, bobcats, corvids, and various snakes were probably the birds' worst natural enemies, in addition to feral dogs and cats. Later, feral hogs became significant predators; coyotes replaced the now-extirpated wolves; and more recently, fire ants have invaded the entire remaining range of the Attwater's prairie-chicken.

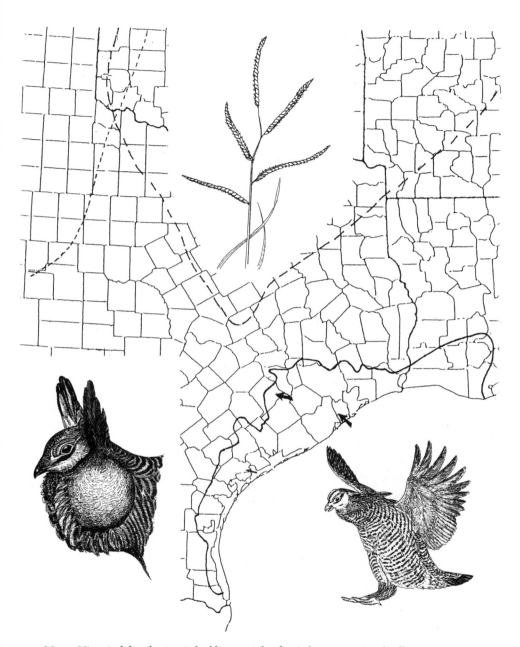

Map 3. Historical distribution (inked line; mainly after Lehmann 1941) and still-surviving population locations (large arrows) of the Attwater's prairie-chicken. The small arrow indicates a population in Refugio County that survived into the late 1990s. The approximate historical Texas limits of the interior greater (long dashes) and lesser (short dashes) prairie-chickens are also shown. The upper inset shows paspalum grass, an important food plant.

Snakes of several species are now believed to be especially serious nest and egg predators.

During winter, adequate food and cover are most likely to be important limiting factors for the birds, according to Lehmann. In spring, access to suitable display sites, namely areas of short grass from 0.5 to 10 acres, surrounded by heavier grassy cover, are needed. Such areas often occur over alkaline-rich hardpan soils, where plant diversity is low and vegetational growth rates are very limited. Weather during the incubation and hatching periods is of critical importance to prairie-chickens if the young are to survive their highly vulnerable period of early posthatching life. During the hot summers, a source of shade and a supply of water are critical, as is good nesting and brooding cover. Ideal nest sites are those that provide visual cover in well-drained areas. According to Lehmann, nest sites should also be within five yards or so of an opening in the cover, providing a possible escape route for the adult should it be threatened on the nest.

SOCIAL AND SEXUAL BEHAVIOR

Of the several hundred display sites observed by Lehmann throughout the study, only one was located on a small knoll. The rest were level with, or even slightly below, the surrounding land. This observation is rather surprising, given the preference of interior prairie-chickens for using elevated sites as display grounds.

Male territorial advertisements begin earlier in Texas than those of the heath hen and northern prairie-chicken. Lehmann reported that by late January or early February the males began to assemble on traditional display sites. During two years, courtship reached its peak in March but continued throughout April. Also in two years, the last display calls were heard on May 20 and May 21. Each morning the birds would display from sunrise until about 8:00, and again in the afternoon from 5:30 until dark.

During Lehmann's study, he counted birds at 5 different display or booming grounds that were used during every spring over his 3-year study period. Average numbers of attending males at these sites ranged from a low mean of 6.8 birds at one site to a maximum mean of 8.7 males. For all 3 years, and considering all 5 sites, the mean number of participating males was 7.7 individuals. This seems a rather low mean participation rate, at least by comparison with that typical of northern prairie-chickens, and probably simply reflects a generally low population density. Only these 5 display grounds, out of the 10 that were under regular observation, were active during all 3 years of Lehmann's study. Of the remainder, one was used only during a single season. This seeming lack of fidelity to display sites is also somewhat surprising and is perhaps the mark of a dispersed and possibly declining population.

In a later study, Maria Dimare examined the possible influence of display ground shape on breeding activity and male mating success. She compared circular-shaped display groups with those of more linear shape and reported that the shape of the group did not influence the total number of males present or the number of males that successfully copulated. However, such "copulatory males" could be distinguished from less successful males by the fact that the overall frequency of their booming calls effectively separated these two groups of males. In a study done about 40 years after Lehmann's, John Horkel also found a high level of variability in booming grounds, especially as to their cover or substrate composition, the numbers of males present, and the numbers of visiting females. On his 6,100-hectare (23-sq-mi) study area there were 27 leks, or somewhat more than 1 lek per square mile. From 1 to 23 males were seen at these leks, and from 0 to 13 females, with the larger leks (13 or more males present) attracting the most females. Of these 27 display sites, 23 were located on artificially maintained substrates, such as gravel or asphalt surfaces associated with roads, oil pads, and oil pipeline rights-of-way; only 4 were located in naturally occurring short grass. It would seem that Texas has indeed paved prairie-chicken paradise and put up a parking lot. One possible disadvantage of the more linear-shaped leks is that they result in instability in the dominance hierarchy of males, possibly allowing less fit males to breed and degrading the genetic fitness of the population. Ten of 11 successful copulations were observed on circular rather than linear leks, although circular leks were seemingly less effective in attracting females than were linear ones.

Judging from Roger Sharpe's studies, there are only minor differences between the booming displays and associated calls of Attwater's and the interior greater prairie-chicken. In both, the mean frequency of the call is about 300 cycles per second (Hz), and each has three distinct syllables, lasting an average of 1.93 seconds in the case of Attwater's. The call is preceded by three rapid tail-snaps and followed by a slower tail-fan. Additionally, the postures and calls associated with territorial defense and associated threats or fights between males are essentially identical (Figure 4).

REPRODUCTIVE BIOLOGY

During Lehmann's study, he was able to locate 19 nests. All were in dead grass of the previous year, and all but two were in tallgrass pastures, the exceptions being in a hay meadow and a fallow field. Favorite nesting materials consisted of dried grasses, namely little bluestem and a paspalum. Fifteen nests were on

Figure 4. Attwater's prairie-chicken, males in territorial confrontation (after U.S. Fish and Wildlife Service photo). Sonogram of cackling vocalization after Sharpe (1968); duration 1.0 second.

well-drained mounds or ridges, and 12 were within 10 yards of a trail, such as a cattle trail. All were also within half a mile of a display ground. Nests with eggs were found as early as February 25 and as late as May 29, the last being a just-completed clutch. The peak period of nesting was late March and early April in Colorado County. The usual clutch size of first nesting attempts was 12 eggs.

Since incubation lasts about 23–24 days, the last chicks should be out near the end of June. At least three hens renested after the loss of their first nests, one of them three times. Such nests were always placed close to the original nest site. However, replacement clutches were smaller; the nests were less well concealed; nest losses were higher; and brood sizes of successfully hatched clutches were smaller. Overall nest losses during two years were high; of 19 nests, only 6 (31 percent) hatched. Brood size averaged 5.48 chicks, who were able to fly short distances at two weeks of age, and at three weeks could fly 40 yards or more.

In John Horkel's study performed during the late 1970s, 19 nests were also found. Nearly half (42 percent) of these nests hatched; 5 percent were abandoned; and predators took 53 percent. All the nests were in midgrass cover, and most were in clumped midgrasses, such as little bluestem. Nests located near rights-of-way had a higher chance of being destroyed by predators, and later nests were less likely to succeed than nests laid early in the season. Throughout the year, the birds were found to prefer using clumped midgrass habitats over all other available cover types, spending an estimated 93 percent of their time in it; they avoided vegetation cover much higher than about two feet tall. Encroachment of brush into midgrass cover also made it less attractive to the birds.

A nesting study by R. Lutz and others in the 1990s used 71 radio-marked females. In the sample of 63 nests found over 5 years, the nesting success (percent of nests hatching eggs successfully) ranged widely, from 19 to 64 percent annually, with a lower percent (0–51) success for renesting birds. A high mortality rate occurred among the radio-equipped females (64 percent) during the nesting season. This result might have been biased owing to the increased risk caused by handling the birds and the physical effects of the transmitters themselves. In a related study, Michael Morrow followed 49 radio-equipped prairie-chickens for three years. During that period, the annual adult survival rates were estimated at 10.8–35.5 percent; nest success averaged 35 percent; and brood survival for the first eight weeks after hatching averaged 34 percent. Of 26 nests found during the study, 85 percent were in fields that had not been burned for three or more years. Young broods also preferred to use fields that had not been burned for at least two years, but older broods moved to sandier areas and vegetational cover associated with first-year burns.

These lines of evidence point to the possibility that the overall reproductive success of the Attwater's prairie-chicken may now be unusually low and might help account for its declining numbers. Markus Peterson looked into this possibility and compared the available data for Attwater's clutch sizes, egg hatchability rates, nesting success rates, brood and chick survivorship rates, and juvenile-to-adult ratios to corresponding figures for northern prairie-chickens. He found that the juvenile-to-adult ratio, nesting success, and the average number of chicks per well-grown brood were all substantially less for the Attwater's than comparable data for northern prairie-chickens. There are obviously many possible influences relative to these observed differences in reproductive success. Weather during the breeding season, parasitism rates, and other factors were all mentioned by Peterson as subjects suitable for investigation. Markus Peterson and Nova Silvy also examined the hypothesis that increased precipitation during May, or throughout the entire March–June period, leads to decreased reproductive success, whereas normal precipitation improves it.

This hypothesis was supported by their available data, although comparable hypotheses related to breeding-season precipitation or spring flooding were not similarly supported.

POPULATION TRENDS AND ECOLOGICAL STUDIES

In 1967 a fall hurricane caused extensive damage and widespread flooding along the Texas coast, and the prairie-chicken population in the most affected part of the bird's range suddenly dropped from about 1,500 to around 250 birds. The total state Attwater's population was then estimated to be at about 1,070 individuals. In that same year the Attwater's prairie-chicken was declared an endangered race. But not until 1973, with the final passage of the Endangered Species Act, did significant federal financing for the preservation of the birds became available.

By 1972 the estimated population was 1,772 birds, an apparent improvement from the late 1960s when the Nature Conservancy had purchased about 3,500 acres of the best remaining habitat, near Eagle Lake in Colorado County. This conservation effort seemingly helped staunch the apparently inexorable decline in the birds' population. The property was transferred to the U.S. Fish and Wildlife Service in 1972, and it became the basis for the Attwater Prairie Chicken National Wildlife Refuge. This refuge has since been enlarged to include nearly 8,000 acres of prairies, cropland, wetlands, and woodlands. A part of Aransas National Wildlife Refuge has also been designated specifically for management as prairie-chicken habitat, and the resulting total federal acreage managed for prairie-chickens now totals some 15,000 noncontiguous acres, a tiny fraction of 1 percent of the birds' original range.

The prairie-chicken population numbered only 926 birds by 1988, or about half the numbers typical of the early 1970s, and by 1989 the entire population of prairie-chickens had again dropped by half, to an estimated 432 birds. These birds were in four small, separate flocks spread across nine counties. About two-thirds of the total population was then present on private ranchlands in Refugio and Goliad counties, reflecting the importance of private landowners in determining the probable fate of the birds. In 1992, when 456 birds were known to be present in the wild, 49 eggs were taken from the wild. From these eggs 35 chicks resulted, after being incubated at Fossil Rim Wildlife Center, near Glen Rose, Texas. These young birds marked the start of a captive breeding program that eventually involved four zoos and Texas A&M University. By 1994 the captive-raised birds began producing their own offspring. The birds might help provide a nucleus of new stock for return to places judged suitable for supplementing the wild population or reintroducing them into new areas. But captive-raised birds have a notoriously low survival rate when they are introduced into the wild and cannot be counted upon to reinvigorate the flock.

By 1995 about 160 birds still survived in the wild, as did 42 more in captivity. The stated goal of the national recovery plan for the birds was to establish a genetically viable population of at least 5,000 birds, distributed in three different areas of Texas. The wild population had slipped even farther by 1998, to 56 birds in the wild, plus some 200 in captive breeding programs. The wild population then represented a 96 percent decrease from that of 1980 and was slightly higher than the low count of 42 birds in 1996. The amount of occupied habitat dropped by 94 percent during that same period, from 297,413 acres to 17,784 acres, and birds had become extirpated from Aransas, Austin, Brazoria, Fort Bend, Goliad, Harris, and Victoria counties.

The largest remaining population in 1998 was in Galveston County, where 18 males occupied about 1,730 acres. That population had peaked at 110 birds in 1981 and was supplemented by captive-raised stock in 1996 and 1997. Refugio County, which had supported 726 birds on 7,900 acres in 1980 and a peak population of 838 birds in 1984, had a 1998 population of no more than 12 males. The Colorado County population, which had peaked at 320 birds in 1983, on about 37,000 acres, had dropped to 8 individuals on the Attwater Prairie Chicken National Wildlife Refuge. Its population had been supplemented by 93 captive-raised birds between 1995 and 1997. The 1980 population of 326 birds in adjacent Austin County had been reduced to zero by 1995. As of the spring of 2000, there were about 45 Attwater's left in the wild, including 28 at the 1200-acre Nature Conservancy preserve (Galveston Bay Prairie Preserve) near Texas City in Galveston County and 18 at the Attwater's National Wildlife Refuge near Eagle Lake in Colorado County. The lek at the refuge had 10 males using it sporadically, and the preserve near Texas City had 14 males using it steadily, according to Joel Satore (personal communication). Birds of the Refugio County population were last seen in 1998 and are now believed to have disappeared.

Diseases, as well as inbreeding, increasingly threaten the captive stock as generation follows generation and innate wildness tends to be bred out of the birds. About 200 birds were produced in captivity in 2000. In one release of 25 adult birds trapped and transplanted from one county to another, only 1 survived more than a year, and no offspring were produced. Predators such as coyotes, bobcats, owls, and snakes have taken their toll. More favorably, however, of 120 captive-reared birds released during summer and fall seasons, at least 40 percent survived to the end of the year, and the survivors were reproductively active the next spring. A retrovirus, reticuloendothelial virus (REV), discovered during 1994 in the captive flock has been found fatal to the birds and produced 42 deaths during one study. Hybridization with northern prairie-chickens and artificial manipulation of day lengths have been used to attempt to increase reproductive activity in captives.

In 2001 the population of wild Attwater's prairie-chickens was at the lowest in history, in spite of all the physical efforts and diverse technological fixes tried in the preceding decades. During the spring of 2001 the last active lek at the Attwater's National Wildlife Refuge collapsed, with half of the 10 participating males being killed or dying within a matter of a few weeks. The remaining males were left without any coherent social and territorial structure. According to Joel Satore, the lek near Texas City dropped from 12 to about 8 or 9 participating males within a year, so it too is marginal in terms of having any potential for maintaining both genetic and social integrity. And the birds' preferred native prairie habitats continue to slip quietly into oblivion. Like the heath hen, the Attwater's prairie-chicken may soon take its place in the hall of fame of birds that our society has failed to save because we did not act in time.

3

FADING FOOTPRINTS IN THE SAND
The Lesser Prairie-Chicken and the Sand Shinnery Grasslands

Texas is a state already well known as a place for making final, futile last stands. So it was at San Antonio for the defenders of the Alamo. So it is now on the coastal plain for the Attwater's prairie-chicken. So it may likely to be in north Texas for the lesser prairie-chicken.

Far beyond the east Texas coastal plain, to the west of the Edwards Plateau of central Texas, and among the upper, often dry, tributaries of the Brazos and Red rivers, lies a little-visited and even less valued part of north Texas. This remote panhandle region is largely marked by sandy plains, alkaline and often temporary lakes, arid and sun-baked flats, and vegetation that is more easily described by its absence than by its profusion or diversity. Here and there are scattered, often slowly vanishing towns and villages with names like "Cactus" and "Shallowater." This portion of Texas and adjacent New Mexico encompasses the so-called Staked Plains, a desolate and arid region so flat and featureless that early Spanish explorers were reputed to have placed tall stakes at intervals along their route so as not to become lost for lack of obvious landmarks. But it was ideal habitat for lesser prairie-chickens.

This region was once largely covered by drought-tolerant perennial grasses, such as several grama grasses (*Bouteloua* spp.) and bluestems, especially little bluestem *(Schizachyrium scoparius)*. Sand dropseed *(Sporobolus cryptandrus)*, sand lovegrass *(Eragrostis trichoides)*, three-awn grass (*Aristida* spp.), and needle-and-thread *(Stipa comata)* were also common, the first two on sandier sites. Shrubs, including soapweed yucca *(Yucca glauca),* also occurred widely, and wild plum (*Prunus* spp.) and aromatic sumac *(Rhus aromatica)* were present on somewhat moister or less sandy sites. Throughout the entire region, sandsage *(Artemisia filifolia)* was the most prevalent shrub on highly sandy soils.

Sandsage is a woody, long-lived, and aromatic shrub that grows to a height of about three feet and is usually much smaller than the more familiar big sagebrush, which is also locally common on firmer ground. Sandsage grows best on well-drained, sandy soil and is perhaps more tolerant of a stabilized dune substrate than are any of the dozens of other species of sage occurring throughout the arid American West, a region that is traditionally regarded as almost synonymous with sagebrush. Sandsage extends north to westernmost South Dakota, south to northern Mexico (see Map 5), and west to the Great Basin region of Utah and Arizona. It has long, narrow, aromatic leaves and a bitter taste, and it is rich in oils that tend to repel leaf-eating insects. It represents a potential livestock food described by range management experts as "poor to worthless" for cattle and "poor to fair" for sheep and horses. Nevertheless, deer and especially antelope can thrive on it. Although sage leaves and flower heads are consumed by lesser prairie-chickens, the plant is not known to be a significant food source. The leaves may, however, provide a substitute for free water in a generally arid habitat. On hot days the birds often rest in its shade; they may also roost among sage and sometimes nest under clumps of sage. Some leaves may persist into the winter, potentially providing emergency foods. The small but numerous seeds of sandsage are known to be a minor food source for the sharp-tailed grouse and may be so for the lesser prairie-chicken.

From western Oklahoma southward, clumps of scrubby oaks, especially shinnery oak *(Quercus havardii),* become increasingly frequent on very sandy soils, where they often share dominance with sandsage and perennial native grasses such as sand dropseed and little bluestem. Shinnery oak is part of that great genus of magnificent American trees with which our country was, quite literally, constructed. Yet shinnery oak not only barely qualifies as a tree, it sometimes doesn't even seem to enter the contest. It is usually nothing more than an inconspicuous shrub, growing up to no more than three feet tall, sometimes barely reaching beyond one's shins. ("Shinnery," however, comes from the Louisiana French *chenière,* meaning an oak woodland.) In spite of a life span of only 11 to 15 years for single aboveground shoots, the underground stems persist through repeated cloning and stem rejuvenation, thus individual plants may reach hundreds if not thousands of years of age. The underground stems may spread to cover an acre or more and form dense mottes that provide escape cover and reliable food sources for many animal species. The tree's usual height is shrublike in sandy, well-drained areas, but in somewhat moister soils, and with the greatest good fortune and freedom from range fires, it may at times attain the stature of a small tree.

Shinnery oak produces a considerable number of small (dime-size) acorns, which are a primary food source during fall and winter for prairie-chickens and are also eaten by them during spring and summer. Its nutritious leaves are easily accessible to smaller mammals and birds, as are its catkins and insect (cynipid wasp) galls, and all these are seasonally consumed by prairie-chickens. The catkins and buds are unusually high in protein content (19–22 percent) and may provide important spring foods. The birds also use shinnery oak as a source of summer shade, overhead nest cover, and perhaps as escape cover or for nocturnal roosting when higher trees are unavailable. Where sand sage communities without oaks occur in the vicinity of those with oaks, prairie-chicken densities are considerably greater in the latter, indicating the ecological importance of this species to the birds.

The oak's 5- to 7-million-acre range (see Map 5), extending from the western half of Oklahoma and the panhandle of Texas plus adjacent parts of eastern New Mexico, almost perfectly circumscribes the primary range of the lesser prairie-chicken. In a curious way, shinnery oak seems to be the lesser prairie-chicken's single closest ecological partner, in the same way that the bear oak was once the ecological counterpart to the heath hen. In recent years, elimination of shinnery oak by defoliating herbicides (especially tebuthiuron) has been a serious problem. This is especially true on federally owned Bureau of Land Management (BLM) lands. That agency abrogated a 1960s understanding with the New Mexico Department of Game and Fish and treated some 100,000 acres of grazing lands with herbicides. Most follow-up work indicated negative effects on lesser prairie-chicken populations and harmful effects on mule deer and lagomorphs, but higher subsequent populations of rodents. In later years, as the BLM began to realize the wildlife benefits of maintaining shinnery oak communities, the agency has restricted cattle grazing and other harmful influences on native wildlife.

The largest publicly owned shinnery oak habitat occurs on BLM lands in New Mexico, where the agency controls 500,000 to 1 million acres. There are also about 500,000 acres of state trust land in New Mexico. In Texas and Oklahoma nearly all the shinnery oak is on privately owned lands. Most of these native scrubby grasslands were long since converted to rangeland for cattle-raising and then to irrigated farmlands, exploiting the underground but easily accessible Ogallala aquifer. The shrubby oaks and sandsage, traditionally favorite lesser prairie-chicken habitats, have largely been eliminated by burning or by defoliation treatments to make room for plants that cattle consider more palatable and humans find more profitable. The aquifer itself has progressively shrunk, now providing less than half the volume of water it offered only 50 years ago, forcing farmers to give up on irrigation-based crops and return to dry-land wheat and sorghum crops.

There is no fossil evidence that provides us with a clue as to the origin of the lesser prairie-chicken. As recently as the mid-20th century it had a range so close to the northern race of greater prairie-chicken of the more mesic grasslands that one wonders if the two might have been in extensive geographic contact during historical times. The lesser prairie-chicken is also only poorly separated geographically from the Attwater's race of prairie-chicken. Some apparent wintering lessers once strayed east almost to the coastal plain of Texas, and early records also suggest there was perhaps a general movement southward in winter, followed by a northward shift in spring. In the early 1920s a few lessers were even collected as far north as southwestern Nebraska, not far from where the interior greater prairie-chickens are still occasionally to be found (Map 4).

Perhaps the ancestral lessers ranged farther south during late Pleistocene times, possibly becoming isolated in the arid Mexican highlands. They then probably remained there long enough to develop the paler plumage, smaller size, and some male display and call differences that now help distinguish them from greater prairie-chickens. Some of these behavioral differences, such as speed of movements and tone frequencies of vocalizations, are the obvious result of body downsizing, but others are less predictable and more distinctive than might have been anticipated as a result of chance evolutionary divergence. Most taxonomists, after looking at these several distinctive features, have concluded that the lesser prairie-chicken should be recognized as a species distinct from all three of the larger prairie-chickens, the now-extinct heath hen and the Attwater's and northern races of the greater prairie-chicken. A few others have disagreed, claiming that all four should simply be called "pinnated grouse," thereby distinguishing them, at least at the species level, from the sharp-tailed grouse and all other grassland grouse of the world.

Although lesser and interior greater prairie-chickens are not now in significant geographic contact and thus have few or no opportunities to interbreed, captive-produced hybrids are fully fertile and appear to be as sexually active as their parental types. These same traits of hybrid fertility and sexual activity are also true of interior greater prairie-chickens and sharp-tailed grouse hybrids (see Figure 16, p. 96). These two grouse sometimes naturally interbreed in the few areas where they are in geographic contact, but the hybridization frequency is evidently now too low to jeopardize their respective gene pools.

There are no good estimates of the original numbers of lesser prairie-chickens in the Southwest and only a few educated guesses. Its maximum historical range, from west-central Texas to west-central Kansas, may have approximated 100,000–150,000 square miles, of which about two-thirds were in Texas. The rest of the range was about equally divided between Oklahoma and Kansas, with New Mexico and Colorado getting the leftovers and Nebraska a few remaining crumbs. One early (1945) estimate by the Texas Game, Fish, and Oys-

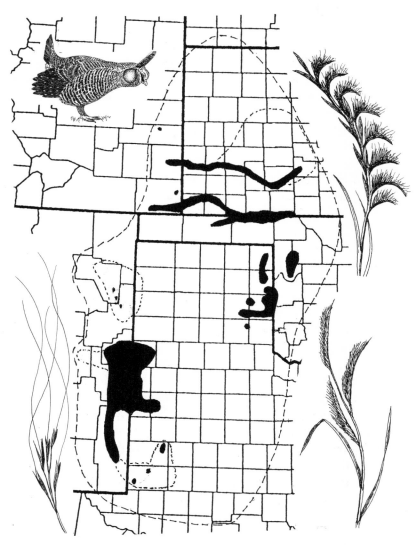

Map 4. Historical (dashed line) and current (inked) distributions of the lesser prairie-chicken. Short dashes enclose a few extirpated or nearly extirpated populations. Inset sketches illustrate needle-and-thread *(left)*, little bluestem *(above right)*, and hairy grama *(below right)*, important native cover grasses of the south-central plains.

ter Commission suggested that as many as 2 million lesser-prairie-chickens may have been present in Texas before 1900, a density representing about 20 birds per square mile. If that is the case, then the overall population of the species might have once approached 3 million. There are no estimates for the original populations of Kansas, Oklahoma, New Mexico, or Colorado. By the mid-1900s the species' total population probably comprised no more than 40,000 birds, or about 1 percent of the suggested 3 million.

As with the Attwater's and interior greater prairie-chickens, no thought was given in Texas to possible conservation of these birds before 1900. Indeed, hunting contests were held in the panhandle as well as along the Gulf Coast, each with up to 50 hunters participating and with the birds often being left to rot where they had been killed. Market hunting was unrestricted and, most important perhaps, the land was rapidly being converted to grazing and agricultural purposes. Probably by 1900 the lesser prairie-chicken populations of Texas had already entered a steep decline, and the tapping of the Ogallala aquifer in north Texas opened a large and previously undeveloped region to cotton growing and small-grain agriculture. Moreover, periodic droughts, especially those of the 1930s, were sometimes devastating both to human and prairie-chicken populations in north Texas. By 1937 the lesser's Texas population was judged to be only about 12,000 birds, or less than 1 percent of its presumed historical status. The Texas state legislature then established its first closed season on prairie-chickens, which in the case of the lesser was to remain in effect until 1967, when a two-day season was initiated. Since that time, two-day seasons have generally been allowed in both the northeastern and southwestern parts of the panhandle. In 1940 the northeastern part of the Texas panhandle had an estimated 1,715 square miles of suitable prairie-chicken habitat and the southwestern region 3,560 square miles, or a total of 5,275 square miles. In 1967 the state's population was estimated at only 10,000 birds, or about 2 birds per square mile. By 1989 these range estimates had been reduced more than 60 percent, to 1,182 and 1,078 square miles, respectively. At a modest 2 birds per square mile, the total Texas population might then have numbered about 4,000 individuals.

Robert Sullivan and coauthors have estimated that the Texas range of lesser prairie-chickens was reduced 78 percent between 1963 and 1980, the losses being particularly great in the southwestern and east-central panhandle, whereas those in the northeastern panhandle remained fairly stable. Mean lek numbers declined precipitously in the southwestern panhandle during this period; during the 1990s they were about 55 percent below the 1969–1989 average. During the same period in the northeastern panhandle they were only about 7 percent below the 1942–1989 average. In the late 1990s small range expansions occurred in Bailey, Cochran, Gray, Hemphill, Lipscomb, Terry, and Wheeler counties as a result of special conservation efforts. An estimate in 2001 by the Texas Game and Parks Department of the panhandle's population of lesser prairie-chickens was 3,000 birds, with most of them in Hemphill, Wheeler, and Lipscomb counties, in the northeastern corner of the panhandle.

In Hemphill County there are some 100,00 acres of sandsage grassland left and in adjoining Wheeler County about 6,720 acres of shinnery oak grassland. However, according to Kenneth Seyffert the density of birds in Hemphill County went from 2,747 acres (1,112 ha) per lek during the period 1967–1987 to 3,317 acres (1,343 ha) per lek between 1986 and 1999, suggesting a population decline of about 20 percent in a few decades. The decline in Wheeler County was even greater during that period, from 425 acres (172 ha) per lek to 5,689 acres (2,303 ha), suggesting a density reduction of about 75 percent. Besides these birds in the northeastern panhandle, there are also some surviving in Bailey, Cochran, and Yoakum counties, along the New Mexico border, and perhaps in nearby Lamb and Andrews counties. Some have also been reported in Hockley, Oldham, and Deaf Smith counties, and seemingly suitable sandsage habitat still exists in Hartley County. Some birds might also still occur in Donley and Collingsworth counties in the northeast, but they seem to be gone from their historical ranges in Armstrong, Carson, Moore, Ochiltree, Parmer, Potter, and Roberts counties. Unpublished data from the Texas Breeding Bird Atlas Web site (http://tbba.cbi.tamucc.edu/) indicate that there were four possible, five probable, and two confirmed Texas breedings between the spring of 1987 and early 1992. Five of the records were from the southwestern panhandle; seven were from the northeastern panhandle.

More than 3 million acres of Texas land have been converted to noncropland cover as part of the Conservation Reserve Program, but much of this acreage is planted to near-monocultures of nonnative grass species apparently unattractive to prairie-chickens. Such vegetation may actually favor potentially significant predator species such as coyotes more than prairie-chickens by providing an increase in suitable cover for the predators. Many of the relict prairies now exist as remnant patches of less than 250 acres, an area far too small to support prairie-chickens. In the last 10 years of the 20th century the population of seven of the counties where lesser prairie-chickens still occur dropped an average of about 1 percent per year, reflecting the long-term human population drain in the region. Additionally, the level of the Ogallala aquifer, the lifeblood for agriculture in the panhandle, dropped more than one foot per year between 1991 and 1996, and during the entire decade of the 1990s the rate of decline averaged slightly under a foot per year. The annual use in Texas represents about 1 percent of the likely total Texas water reserves. The Texas high plains region uses nearly 90 percent of all the water pumped out of the Ogallala aquifer in the state, which is used mostly for center pivot irrigation. When the Texas aquifer finally runs dry, perhaps in less than a hundred years, the land might revert to prairie-chicken habitat even if no prairie-chickens are there to reclaim it.

The exact status of New Mexico's lesser prairie-chicken population has long been something of a mystery. The state's Department of Game and Fish has undertaken few rangewide studies of its distribution and only in the late 1990s began to conduct extensive display-ground counts for estimating population trends. At the start of the twentieth century the lesser probably still ranged rather widely over at least eight eastern counties, from Union and Harding in the north to Lea and Eddy in the south. Locally it extended west to the Pecos River Valley, and its total range may have included about 15,000 square miles. In 1968 James Sands reported that the largest remaining New Mexico populations were in Roosevelt and northern Lea counties, with a few also in eastern Chaves and parts of De Baca, Quay, and Curry counties. During the early 1970s the state's population was thought by Sands to be about 8,000 to 10,000 birds, down substantially from the 40,000 to 50,000 of the 1950s. Survey data for the period 1971 to 1997 analyzed by the New Mexico Heritage Institute indicated that a clear population decline occurred after 1988. In the post–World War II years the lesser prairie-chicken remained legal game in New Mexico, at least until 1996. The average annual harvest during the 1960s was about 1,000 birds, but by 1979 was reduced to only about 130. Maximum hunter harvests of about 4,000 birds occurred in 1987 and 1988, but these numbers subsequently declined rapidly and the season was finally closed in 1996.

James Bailey and Sartor Williams have estimated that the species' New Mexico population was once about 125,000 birds, occupying some 38,000 square kilometers (14,672 sq mi) a mean density of about 3.3 birds per square kilometer (8.5 birds per sq mi). By 1961 the population was estimated at 40,000 to 50,000 birds, and in 1968 at 8,000 to 10,000 birds. In 1979 the estimate was still 10,000. By the late 1990s the lesser prairie-chicken's historical range had been reduced by more than half, with most of the remaining birds occurring in about 20 percent of the original range, mainly on privately owned lands in southern Roosevelt, extreme northern Lea, and eastern Chaves counties. The area shown on Map 5 as fully occupied in northern Roosevelt and Curry counties now actually consists of only sparse and isolated populations, according to Bailey and Williams. In west-central Lea County, where there were 20 leks in 1987, only a single lek was found in 2000. The birds now appear to be entirely gone from their historical range within Union, Harding, and Quay counties in northeastern New Mexico, and nearly all those still surviving as late as 1997–1998 in east-central New Mexico occurred within about 25 miles of the Texas border, near Portales. On the Texas side of this border some birds were still present in Bailey, Cochran, and Yoakum counties during the late 1990s.

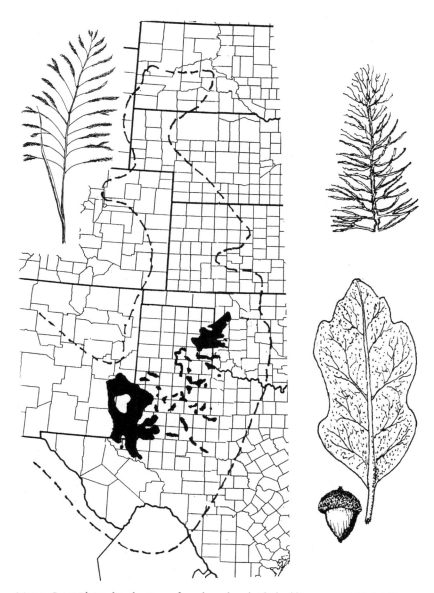

Map 5. Great Plains distributions of sand sagebrush (dashed line; *upper right sketch*) and shinnery oak community (inked; *lower right sketch*), important native cover and food plants for the lesser prairie-chicken. Also shown *(upper left)* is sand dropseed, a common native grass of sandy soils in the region. Distribution of shinnery oak community after Peterson and Boyd (1998).

Two years of drought, starting in 1989, resulted in an abrupt population decline throughout the New Mexican range, a trend that may have been exacerbated by excessive grazing, development of gas and oil reserves, the control of so-designated undesirable native shrubs, especially sand sagebrush and shinnery oak, and the inherent genetic and ecological problems associated with small, isolated populations. Shrub control on BLM land has contributed to this problem, but uncontrolled herbicide applications on private lands have probably had greater negative effects on native habitats. Roadside lek surveys in the late 1990s revealed up to 53 leks in east-central New Mexico, but counting method variations have not permitted reliable population trends to be established. Only 2 leks were found in southeastern New Mexico in 1999, and none were found that year in northeastern New Mexico. Assuming a mean of about 8 males per lek, the spring 2000 population in New Mexico might well have been under 500 males, or a total of about 1,000 breeders.

In 1997 the lesser prairie-chicken was described as imperiled in the state by the state's Natural Heritage Program. In October of that year the New Mexico Department of Game and Fish announced that it would begin a study to determine whether the species should be officially listed as threatened or endangered under the provisions of the state's Wildlife Conservation Act. However, two years later, the New Mexico Game Commission refused to accept the recommendation of its departmental director to list the species as state-threatened, evidently largely as a result of pressures from New Mexico's Cattle Growers Association. Instead, the commission announced an "interim management approach," a temporary management plan intended to last for at least six years, thus effectively delaying the listing process.

The story in Oklahoma is similarly discouraging. At one time most of the state lying west of the 100th meridian was probably occupied by lesser prairie-chickens, as it essentially comprised arid prairies. As elsewhere, the birds were most abundant on sites with sandy soils and some brushy plants, usually sand sagebrush and shinnery oak. This oak usually grows only a few feet high, but occasionally dense growths of much taller plants form mottes that may be several hundred feet in diameter. The trees rarely reach 20 feet in height, these tall trees actually being hybrids with post oak *(Q. stellata)*. On mixed-grass prairies with firmer soils, sand sage is less evident, and the brushy component is made up largely of skunkbrush sumac and wild plum. Such habitats approach those of the interior greater prairie-chicken, and in Oklahoma, as in Texas and Kansas, the breeding ranges of these two species approached one another closely. The lesser certainly ranged east at least to Woods and Woodward counties, and probably to Major, as recently as the 1940s, whereas interior greater prairie-chickens were then present as far west as Kay and Noble counties, only about 50 miles to the east.

No early estimates of lesser prairie-chicken populations exist for Oklahoma. The birds were legally hunted with few restrictions up until 1915, but thereafter the season was opened only periodically until 1951. Based on a few sample counts by state Wildlife Conservation Department biologists in 1940 from study areas totaling only 20 square miles, a statewide population of 14,914 birds was extrapolated. Such estimations obviously have little value as such but are often the only numbers available. Protection continued throughout the 1940s, but short, one- and three-day hunting seasons were allowed in 1950 and 1951. The species' state range seemingly retracted hardly at all during the 1940s and 1950s, although there were severe droughts between 1952 and 1955. The latter half of the 1950s were more favorable for breeding, and in 1956 average density estimates of 6.5 males per square mile were determined for mixed-grass prairie; 4.0 males, in shinnery oak; and 1.75 males per square mile, in sandsage grasslands.

By 1960 the species' known range in Oklahoma was calculated at about 2,400 square miles, and another 1,400 square miles of potential but unoccupied grasslands range existed. The estimated spring population of males was 15,000, the same total as 1940. However, estimated spring densities were higher than in 1956, as many as 11.25 males estimated per square mile in shinnery oak habitats and as low as 2.3 in sandsage grasslands. In one study area the density of males increased progressively from 1956 to 1962, but more droughts during the late 1960s and early 1970s caused a population crash. In 1979 the spring male population for the entire state was estimated at about 7,500 birds, or about half as many as had been estimated in 1960, and the occupied range had also declined by 55 percent. More than half the remaining range consisted of sand sagebrush; nearly all the rest was shinnery oak habitat.

From 1980 onward display-ground counts have been performed in several counties. The 11-year (1980–1990) average of males per lek then hovered between 6 and 8 (11-year mean 7.4 males), without any clearly apparent trend. However, between 1988 and 1999 the density index declined about 80 percent, probably as a result of the severe droughts of the 1990s. Hunting of both species of prairie-chickens was terminated in Oklahoma after 1997. In the spring of 2000 lek counts indicated a reduction of 31 percent in the number of males attending leks relative to 1999, down to 4.6 males per lek. The average lek density (12.5 square miles per lek) remained the same.

An analysis of Oklahoma's lesser prairie-chickens by Russell Horton suggested that the species' range has diminished by about 64 percent according to historic records and now includes only 8 of the 22 counties where they once occurred. As of the year 2000, fewer than 3,000 birds were present during the breeding season, or 20 percent of the estimated 1940 population. A landscape level analysis of prairie-chicken habitats in Oklahoma, New Mexico, and Texas revealed that of 13 studied lesser prairie-chicken populations, 5 were judged as declining; 4 of these were in Oklahoma. In all regions, reductions in the total amount of available shrublands, rather than changes in specific cover types, correlated most closely with population losses. In Oklahoma these landscape cover changes were estimated at a rate of 11 percent per decade, as compared with 3 percent in Texas and 1 percent in New Mexico.

Kansas is a prairie state that was once fully occupied by prairie-chickens, the lesser to the west and south, and the northern race of the greater to the east and north. The sandy lands immediately to the south of the Arkansas and Cimarron rivers are the core of the lesser's range in Kansas, but there are winter records well to the north and also to the east of these regions. In the Arkansas River Valley the birds are most numerous from the Colorado line to the general vicinity of the Great Bend region. The apparent eastern limits there (in Edwards County) approach the western limits of the greater prairie-chicken, and the two species evidently have some local contact. Although the prairie-chickens of Quivera National Wildlife Refuge (in Stafford County) are typically greaters, two refuge biologists thought they observed lessers there in the spring of 2001. Matthew Bain, a graduate student at Fort Hays State University, informed me that as of spring 2001 he had found mixed leks in 4 counties north of the Arkansas River, namely Ness, Gove, Trego, and Ellis (personal communication). In 2001 he surveyed 57 leks there, of which 13 were mixed, 16 consisted of lessers only, and 28 were of greaters only. Hybrid males have been observed on 2 leks. His study of species interactions and hybridization is still under way.

Nobody knows how many lessers once occurred in Kansas, but one remarkable estimate of 15,000–20,000 in a single Seward County grain field (unstated size) in 1904 gives some idea of the immense numbers that must once have existed. The drought years of the 1930s nearly doomed the lesser prairie-chickens in Kansas and elsewhere in the southern plains; by one account they were reduced to surviving in Kansas on only two large ranches, in Seward and Meade counties. By the 1950s they had recovered and were known to be present in at least 14 counties, but by then most of their prime habitats had been converted to weedy, abandoned ranchland or to farmland. The advent of irrigation, especially center-pivot irrigation, has effectively spelled the end for the sandsage grassland habitats of western Kansas. When these finally disappear the lesser prairie-chicken will as well.

The lesser prairie-chicken has never been a significant game bird in Kansas, at least relative to the interior greater. As compared with the well-documented greater, relatively little long-term tracking of their populations has been attempted. Hunter kills for the lessers have been estimated annually in Kansas since at least 1975 and have ranged from a high of 6,200 in 1982 to a low of 100 in 1996. Annual average statewide kills were 2,600 for the late 1970s, 2,500 for the 1980s, and 560 for the 1990s through 1998, the last year for which figures are available. Only Texas currently allows the hunting of lesser prairie-chickens, and the Texas kill is relatively small by comparison. Kansas seasons in the 1990s lasted for two months, with a daily limit of a single bird.

Max Thompson and Charles Ely estimated that perhaps 10,000 to 15,000 birds might exist in Kansas in the early 1990s, but the basis for this estimate is unknown. By the spring of 2000, annual lek surveys indicated population increases during each of the previous three years among 10 survey routes south of the Arkansas River. Apparently because of Conservation Reserve Program (CRP) plantings, the population north of the Arkansas River has also increased to a least 165 leks as of the spring of 2001. These trends provide two of the bright spots in an otherwise depressing picture. William Jensen and others have documented the population trends of lesser prairie-chickens in Kansas. They reported that in 2000 the species occupied 31 of the 39 counties representing its historic range, but statewide its population was declining. Although this apparent downward trend may be the result of a statistical artifact, it is more likely the result of habitat loss and deterioration.

Of all the states within the historical range of the lesser prairie-chicken, Colorado has probably always had the fewest birds. Historical records suggest it may once have occurred only in six counties (Baca, Bent, Cheyenne, Kiowa, Lincoln, and Prowers) during presettlement times. It has never been common, and Colorado was the first state to ban its hunting, in the early 1900s. As was the case elsewhere, the dust-bowl days of the 1930s following a period of already marginal farming and ranching on arid lands almost ended the Colorado populations. During that period the species' overall range probably decreased about 92 percent, and its population declined about 97 percent. Display-ground counts by Kenneth Giesen suggested that in the early 1990s Colorado's spring population was at about 1,000 to 2,000 birds. It had increased gradually since the early 1970s, when it was listed as a threatened species in Colorado. Studies by Giesen in the late 1980s and early 1990s indicated that about 35 to 45 display grounds were then known in the state, with an average of 9.6 males present per ground. Colorado's largest remaining population of lesser prairie-chickens is near Campo, Baca County, in extreme southeastern Colorado. Here the Comanche National Grasslands and the Cimarron River provide an extension of the comparable sandsage habitat occurring in the Cimarron National Grassland of adjacent Kansas. There is a smaller population in Prowers County along the Arkansas River, and an even smaller one in Kiowa County along Big Sandy Creek. The species disappeared from Bent County during the early 1940s and has also been extirpated from Lincoln and Cheyenne counties.

Fieldwork done in conjunction with the *Colorado Breeding Bird Atlas* indicated that most breeding-season sightings occurred in shortgrass prairies, with fewer in altered mixed grasses and still fewer in low sagebrush. However, the absence of sand sagebrush on shortgrass prairies dominated only by grama grasses (*Bouteloua* spp.) and buffalo grass *(Buchloe dactyloides)* is known to have a negative effect on lesser prairie-chickens, and Giesen suggested that management plans enhancing the abundance of sandsage may help increase the state's population. Lek counts in the spring of 2000 resulted in a total count of 27 leks and 316 birds, including some females. This result was a substantial improvement over counts the previous spring, perhaps because of CRP plantings. But drought conditions in southeastern Colorado continued to result in poor reproduction there, and future population declines were expected. Kenneth Giesen estimated in 2000 that the total Colorado population then may have numbered fewer than 1,500 breeding birds. Lek counts in 2001 revealed 298 birds on 30 leks, down slightly from the previous year.

Considering the available information from all five states where lesser prairie-chickens still occur, the only thing that can be said with certainty is that the birds have been declining almost everywhere.

SEXUAL BEHAVIOR AND REPRODUCTIVE BIOLOGY

The immediate impression I had on first seeing and hearing lesser prairie-chickens engaged in territorial and courtship displays was that I was watching a fast-forward version of the greater's displays, with a touch of the Marx Brothers thrown in. As compared with greaters, the birds' movements seemed too rapid, and their vocalizations too high-pitched and frenzied; these impressions collectively produced a kind of comic-opera effect.

Like greater prairie-chickens, lesser prairie-chicken males gather in small groups on shortgrass or shrub-laced display grounds, or leks. Likewise, lek locations tend to be permanent, and once males become established on particular territories, they tend to return to the same lek year after year. Males unable to establish territories may form "satellite" leks nearby. If these new locations manage to attract females, the number of leks in an area gradually increases. Although there is seemingly a great deal more running back and forth to defend territorial boundaries in lessers than in greaters, the individual males are similarly spaced out on relatively small territories, the territories having been established and maintained through daily threats and occasional fighting. Fighting by lessers, however, is generally less violent and less prolonged than is the case with greaters. The position, and perhaps the size, of any given territory nevertheless represents a direct reflection of that male's individual vigor, and thus his relative fitness for reproduction. Like the greaters, male lessers concentrate most of their displays during the hour or two surrounding sunrise, over a period of two or three spring months. A moderate amount of display may occur during autumn, and some early evening displays may also occur. All matings, however, are evidently attained during early-morning hours over a period of a few weeks in spring, generally centered in mid-April.

A female visits these leks, probably the nearest available one, for only a few days immediately before the start of egg-laying. Somehow females can recognize which males are most dominant, and each usually selects that particular individual for mating. After being fertilized, females do not return again for mating until the next year unless renesting becomes necessary. These overall similarities between prairie-chickens and other lek-forming grouse tend to end here, for specific, or species-level, differences within broader similarity are also important components of display if the birds are to avoid potential hybridization.

One of the first people to study the display behavior and reproductive biology of the lesser prairie-chicken was Farrell Copelin. He did this work in the late 1950s and early 1960s on a 16-square-mile study tract in Ellis County, Oklahoma. At the time of his study the spring lesser prairie-chicken density in that area ranged from 15 to 20 males per square mile. This is a rather high density for any area, but represents only half the density that had been observed during a comparable study begun before the droughts of the 1930s. Over a five-year period, Copelin found a total of 28 leks on this 16-square-mile area, with from 8 to 21 active in any given year. From 121 to 291 males were present annually, with yearly averages of 14 to 16 males present per lek. The largest leks, with as many as 43 males, seemed to be the ones most consistently used from year to year. Of the 28 leks, 17 were in exactly the same locations as during the 1930s when the area had been last studied. And at least one was used every year during the two studies, these collectively totaling 11 years. Surprisingly, most of these leks were also active during the fall months when older and experienced males reclaimed their old territories, and the youngest ones sometimes simply milled about, probably learning the lek locations and the accepted rules for participating in the spring. A few females visited the leks during fall, but no matings were ever observed during that season. During his studies, Copelin recognized 17 wing-marked and leg-banded birds over a period of two or more seasons. Fifteen of these males occupied the same territory each season. The other 2 remained on the same lek, but altered their territorial positions. Another male moved from one lek to a nearby one nearly a mile away when the lek he had been using was abandoned.

Most of the leks were established on ridges or other elevated sites in shortgrass vegetation; only 1 of 44 was on plowed ground. In sandsage habitats where the ridges were brushy, leks were chosen in shortgrass meadows. Most individual territories were established by mid-April, although some activity on the leks began in late February. These territories observed by Copelin averaged only 12 to 15 feet in diameter, a small size for prairie grouse. In similar leks studied by Ingemar Hjorth in Kansas, he found that the birds were more likely to choose leks on smooth ground than to select for elevated sites, and all the territories there were at least 7 meters (23 ft) in diameter. On a lek I visited in western Kansas, 18 males were gathered within a 30-yard distance, mostly situated along the upper slopes of a gentle dune that was well vegetated with grasses and sandsage up to three feet tall. The birds performed primarily on the grassy substrate between the shrubs but sometimes flew up to perch in a sage where they would look about, cackle, and sometimes even yodel.

Territorial disputes observed by Copelin reached their peak in March and early April. Females visited the leks from the last half of March through the first week in May, with a peak number present during the third week of April. Copulations were seen between April 24 and May 6, with the maximum number (four) observed on April 26. A variety of other studies performed throughout the lesser prairie-chicken's range agree with this timetable, with the second and third weeks of April usually representing the peak period for female attendance and highest copulation frequency. On the lek I observed in the second week of April, at least four females were present during most of the period of intense male activity, and two copulations occurred, plus several other attempted copulations. Although Copelin did not comment on differential mating success in copulation among individual males, Roger Sharpe observed that on one lek a single dominant male obtained 13 of 27 observed copulations (48 percent), while on another lek a single male obtained 11 of 13 (85 percent). This "master cock" mating trait is typical of all lek-forming species and seems to be strongly correlated with relative individual male dominance. Sharpe noticed that the dominant male on the leks he studied consistently drove intruding males out of his territory, readily attacked any other male that he observed copulating, but was himself never interrupted during mating.

Sharpe observed several features of lesser prairie-chicken display that set it apart from both of the two races of greaters. The auditory portion of the male lesser's primary advertisement display lasts only about 0.6 second, rather than 1.96 seconds, and its average pitch is about 500 cycles per second (Hz) higher than in greaters, which is around 800 Hz. Interestingly, in a hybrid male studied by John Crawford, the vocalization had a mean duration intermediate (1.22 seconds) between the two parentals. However, it usually had six distinct syllables rather than the three typical of both parentals. A pair of these captive-bred hybrids proved fertile, producing four second-generation chicks out of 26 eggs laid. Wild hybrids have recently been reported in a few areas of limited geographic contact in central Kansas.

Like the greater's booming, the lesser's equivalent display (usually called "gobbling" or "yodeling") is generally preceded by rapid foot-stamping and followed by an exaggerated tail-spreading as the first of three rapid gobblelike notes is uttered. At this time the head is jerked downward and the lateral reddish throat sacs are inflated (Figures 5 and 6). During the short second note, the head is jerked back upward. The third phrase lasts longest, as is also true in greaters, when the throat sacs deflate and a more normal posture is gradually assumed. The tail is not fanned again during this final stage. The associated call is similar to a small dog's excited barking, but generally has richer, more liquid tones.

Some "low-intensity" gobbling, which lasts only about half as long as the usual type, may also occur, although at least four (rather than three) syllables are quickly uttered, according to Roger Sharpe. Another minor call is the "squeak," the functional counterpart of whooping in greater prairie-chickens; like whooping it is uttered when a female is present on the lek. However, it is much softer than the whoop; acoustically it is more like the "chilk" call of sharp-tailed grouse. Additionally, a rapid wing-shuffling sometimes occurs between bouts of booming, perhaps representing a distinct if minor display.

A common distinctive postural and vocal variation on gobbling is one that Ingemar Hjorth called "bubbling," but "gurgling" might be a better descriptive

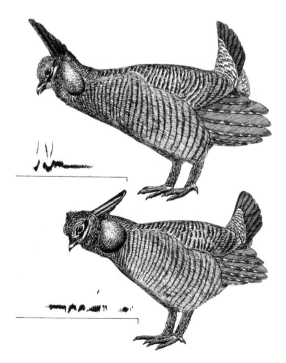

Figure 5. Booming postures of lesser prairie-chicken, with associated vocalizations, including yodeling (*above*) and bubbling (*below*; both after photos by the author). Sonograms after Hjorth (1970); duration 1.0 second.

Figure 6. Lesser prairie-chicken, male booming (after photo by the author).

term. In this display five or six rapid melodic notes are produced, sounding something like the sound of air bubbles rapidly coming out of water. The amount of pinnae erection varies; the tail-swishing is less pronounced but more pulsating; and the primaries are rapidly quivered. The brilliant yellow eye-combs are fully enlarged, and the reddish air sacs are even more fully expanded than during yodeling. The display may precede or end a yodeling sequence but seems especially frequent when two males meet near the edges of their respective territories.

Male lessers do not orient their bubbling/tooting postures toward specific females or toward other males; instead they pivot quickly about between sequences or run a few feet before pausing to repeat their performance. This nearly nonstop activity is almost dizzying to a person used to watching the more sedate, almost magisterial performances typical of greater prairie-chickens. Added to this visual confusion is the virtually constant sound of males yodeling or bubbling simultaneously. These pleasant sounds are frequently interspersed with a cacophony of rapid, high-pitched cackling, similar to hysterical human laughter.

Another feature that distinguishes lessers' displays from that of the greaters, and indeed from all other socially displaying grouse, is the tendency of male lessers to engage in directly competitive simultaneous displays. These consist of rapidly repeated, sequentially overlapping bubbling contests between nearby males. As many as 10 such bubbling sequences may be performed in rapid succession by both birds, their joint actions producing a prolonged and erratically melodic sequence of sounds somewhat resembling the irregular noises made by the bubbling "mudpots" at Yellowstone Park.

Ingemar Hjorth called this interactive bubbling behavior "duetting," noting that the participating males may stand up to 20 feet apart. Often, however, they are much closer, sometimes standing only a foot apart, or nearly within each other's pecking range. The first bird's display initiation is joined by the second with a time-leg averaging only 0.3 second. They then display in concert for several seconds, and stop at about the same time. Such interactive bubbling sequences usually last three to four seconds, but may be much longer. The result is a nearly continuous train of sound that can be heard more than a mile away. Roger Sharpe suggested that such acoustic competition may be the result of the birds' spending more time on the lek than do greaters, on both a seasonal and daily basis, and that this form of ritualized aggression may help reduce the degree of actual fighting among males.

In addition to competitive booming, males on adjoining territories spend a good deal of time threatening one another along territorial boundaries (Figures 7 and 8). Calls uttered at such times include rapid cackles and less intense whining notes. Quick, vertical movements of the head are also common during these encounters. The wings may be held close to the body or partly outstretched, ready for a quick response to an attack. Sometimes the opposing males perform wing-flapping, preening, or pecking movements toward the ground.

In a study of lek locations and dispersion patterns, Brian Locke found that local populations tend to expand by forming new leks rather than by increas-

Figure 7. Lesser prairie-chicken, males in crouching territorial confrontation (after Oklahoma Department of Wildlife Conservation photo). Sonogram of prolonged cackling after Giesen (1998); duration 3.0 seconds.

Figure 8. Lesser prairie-chicken, males in standing territorial confrontation (after photo by Roger Sharpe). Sonograms of cackling *(above)* and whining *(below)* after Hjorth (1970); duration 1.0 second.

ing the average number of males per lek. He noted that the leks were more closely spaced than would be predicted by some lek-spacing hypotheses, but also noted that some apparently suitable lek sites such as oil pads were not used. He believed the birds can hear and respond to adjacent leks from as far away as 1,000 yards. Locke also observed that most leks were situated closer together than the average size of a female's spring home range, which is typically less than a square mile. Because of this lek spacing, females generally have more than one lek they might easily visit. By one hypothesis a female simply chooses the one with the largest number of available males. This tendency results in the so-called male buffet model of lek formation and would tend to produce large lek sizes, as the largest leks would attract the most females. However, the larger the lek size, the lower the chance of any participating male's ever getting a mating opportunity. Alternatively, females might be attracted to specific and highly attractive "hotshot" males, around which other males could perhaps benefit by their simple association. In either case, but especially the latter, this male clustering tendency that produces lekking behavior could occur because nondominant males might occasionally steal a few matings while the master cock is otherwise occupied. Or perhaps by simply living long enough and competing strongly enough, a persistent male might gradually work his way up the dominance ranks, ultimately taking over the position of master cock should he live long enough. Few birds live more than five years in the wild, so it is likely that, as with greater prairie chickens, most master cocks are three to four years of age.

Lek sizes have been studied in a variety of areas, and the mean numbers of males present has usually ranged between 10 and 21. The largest number of males ever reported at a single lek was evidently the 43 seen by Copelin. But among 434 Colorado leks, the maximum observed number of attending males (42) was almost as great. More males are typically present at leks located on native rangeland than on human-altered sites, and in areas of high lek density the density of males present on individual leks is also high, regardless of the total number of males present. Dominant males situated at the center of leks have smaller territories than do peripheral males, indicating that a male's relative female-attracting behavior or his relative territorial position, but not his territorial size, influence individual mating choice by females.

As with other lekking birds, no pair-bonding is associated with courtship. The only individualized behavior is that related to mating itself, which takes but a few seconds. Females may visit more than one lek in the course of a spring display season, but there is no evidence for any of the prairie grouse that more than a single mating is needed to fertilize an entire clutch of a dozen or more eggs. Therefore, unless a nest failure requires the initiation of a new clutch, it is unlikely that a female will revisit the lek following her first successful mating.

After such a mating, the female leaves the lek and apparently heads directly to a nest site. The site is often more than a mile from the lek where she was fertilized and may even be closer to other leks. The first egg is probably laid within a few days after copulation, although some have suggested that a longer period between mating and egg-laying may elapse. Females choose sites that have good concealment features, both vertically and horizontally, with sandsage or shinnery oak often serving this purpose. These shrubs, plus associated grasses and forbs, tend to be of greater density immediately around nest sites than is true of surrounding rangelands. Nests are often placed on slightly sloping land with a north or northeastern exposure, providing some protection from sunlight and hot southwesterly winds.

As with the greater prairie-chickens, incubation lasts 24 to 26 days, is done entirely by the female, and starts with the completion of the clutch. Most clutches have about 10 eggs, rarely as few as 8 or as many as 14. Late-season clutches, and those associated with renesting, average fewer in number. Among a total of 10 different studies whose results were summarized by Kenneth Giesen, an average of only 28 percent of the nests hatched successfully. Predation by various mammals (coyotes and skunks), snakes, and birds (corvids and hawks) is a major cause of egg loss, as is true for most ground-nesting birds. Drought, a late nesting onset, livestock grazing, and reduced nesting cover all negatively influence nesting success, whereas increased height, abundance, and density of native grasses near the nest have favorable influences. The presence of overhead vegetational cover such as shinnery oak may also improve nest success by reducing the nest's visibility to overhead predators or by providing cooling effects in hot weather.

Brood size varies greatly between years and depending on the age of the chicks when counts are made, with broods produced during drought years smaller in number than those in years of better precipitation. By the time fall arrives roughly half the flock should be composed of immature birds, a proportion needed to compensate for the roughly 50 percent annual mortality rate of adults.

CONSERVATION DEVELOPMENTS

In October 1995 the U.S. Fish and Wildlife Service received a formal petition requesting that it list the lesser prairie-chicken as a nationally threatened species. More than a year later (July 1997), the agency finally admitted that enough evidence existed in the petition to warrant a formal status investigation. Since most lesser prairie-chickens occur on privately owned grazing lands or on BLM lands leased for grazing purposes, ranchers were not enamored with the idea of listing the species as legally threatened. In June 1998 the Fish and Wildlife Service neatly skirted the controversy by concluding that such a listing was biologically warranted but was precluded because of higher conservation priorities for other even more seriously threatened species.

After this initial petition for threatened status was submitted, a five-state consortium of conservation agencies formed the Lesser Prairie-Chicken Interstate Working Group to coordinate management activities and evaluate conservation needs. Their 51-page report, an assessment and conservation strategy for the species, was released in February 1999. Partly as a result of their recommendations, about 80,000 acres of private lands in Oklahoma and New Mexico were designated for habitat improvement under "candidate species conservation agreements." Participating ranchers joined the High Plains Partnership for Species at Risk, a consortium of state and federal wildlife agencies, private conservation groups, and private landowners. At best it is an improbable alliance of strange bedfellows, and the results are likely to be unpredictable.

As of the year 2001, the world population of lesser prairie-chickens is apparently somewhere between 10,000 and 20,000 breeding-season birds, with up to 3,000 each in Texas and Oklahoma, about 1,500 in Colorado, fewer than 1,000 in New Mexico, and an undetermined number in Kansas, perhaps 5,000 to 10,000. Kansas, therefore, must represent our last, best chance of saving the species from extinction. A population of 10,000 to 20,000 birds may seem to be a comfortable number, but the history of the Attwater's prairie-chicken should provide ample warning that this may not be the case, as the lesser prairie-chicken is in apparent decline almost everywhere.

4

A DRUMMING AT FIRST LIGHT
The Interior Greater Prairie-Chicken and the Tallgrass Prairies

Late March sunrises on the glacial-shaped plains of southeastern Nebraska are unlike those of any other place in the world. At least so it seems to anyone lucky enough to experience one while sitting inside a canvas-covered blind with frost-covered native grasses at one's feet and a sky tinted with pink clouds to the east and flecked with fading stars westward. In a nearby grove great horned owls occasionally complain softly of the end of another night's hunt; in the distance a coyote utters its last yipping notes. Then all is quiet for a time as the earth slowly proceeds into the future, turning present into past and transforming reality into memory.

A whispering of wings overhead signals that the first prairie-chickens may be arriving at the lek. I look out of a peephole into the gloom. The hillside seems to be deserted, and quiet returns. Then, ever so softly comes a low, half-humming, half-drumming sound. It seems to emanate more from my imagination than from the real world, evoking images of long-gone bison herds, of a time now so remote it is more myth than history. Peering out once again into the near-darkness, I discern the silhouette of a strangely shaped creature, less grouse than ghost, along the horizon's crest. The mysterious humming, now louder and more confident sounding, penetrates the blind. It has a calming effect, resembling mantra more than melody, magic more than reality. Soon the call is answered by other nearly invisible players, some walking in, others flying in. Before long, continuous sound surrounds and penetrates my body and soul. The eastern sky is now flushed with red, and the glory of another prairie sunrise is under way.

There was a time during the middle to late 1800s when most people living across the unforested areas of central North America could have experienced such a morning as I have described. The interior greater prairie-chicken had a vast historical range that encompassed the tallgrass prairies of the Central Plains and extended from central Texas to eastern North Dakota, and east to Tennessee, Kentucky, and Ohio (Map 6). It thus roughly circumscribed the original range of North America's tallgrass prairies.

The easternmost populations of interior greater prairie-chickens were the first to be influenced by ecological changes brought on by European agriculture, at about the same time the heath hen was disappearing from the Atlantic seaboard. By the early 1800s the Kentucky population of prairie-chickens was probably declining, and by 1850 the Tennessee population had vanished. It was gone in Kentucky by 1874, and in Arkansas by 1913. We have no way of knowing how large any of these populations once were, but all occurred east of the main region of the tallgrass prairies. Here the birds were probably confined to grassy islands in and around the margins of the extensive hardwood forest, such as the barrens of southwestern Kentucky. There an abundance of nearby oaks provided winter foods, in a manner similar to the food sources utilized by heath hen populations in the east.

Farther west, in Texas, the birds were being slaughtered with abandon. A population that numbered an estimated half a million birds in 1850 was shot by the wagonload for market or killed simply for sport. The fertile prairies of northeastern Texas, their prime habitat, were being plowed under as rapidly as humanly possible. The last flock was seen in Harrison County in 1920. After that time there were only scattered sightings of individual birds in Texas, some perhaps strays from Oklahoma, and one was seen as late as the 1950s. Concurrently the Ohio population flickered out, the last record occurring about 1934.

Toward the end of the 19th century, as the birds were disappearing in the east and south, they were increasing and expanding in the north and west, both in the Dakotas and in the southern parts of Canada's prairie provinces. With the development of small-grain agriculture in these northern mixed-grass and tallgrass prairies, the prairie-chicken encountered an ideal combination of native grasses for nesting cover, an abundant source of invertebrate foods for rearing young, and a superabundance of small grains such as wheat to tide them over the fall and winter months. This was especially noticeable in southern Canada, where the prairie-chicken rather rapidly extended its range more than 500 miles northwest to east-central Alberta. It rapidly colonized these regions during the last two decades of the 1800s and then declined with equal rapidity during the first two decades of the 20th century. It had almost completely disappeared during the droughts of the 1930s but persisted in very small numbers into the 1960s in Alberta and into the 1970s in Saskatchewan and Manitoba.

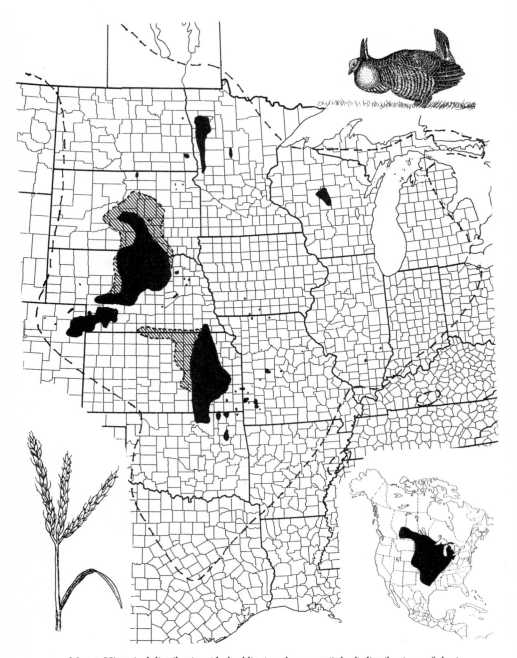

Map 6. Historical distribution (dashed line) and current (inked) distributions of the interior greater prairie-chicken. Marginal low-density populations in the primary range from South Dakota through Kansas are also shown (hatching). The race's maximum historical distribution is shown in the inset map *(lower right);* the inset sketch *(lower left)* illustrates big bluestem, an important tallgrass cover species.

The northern extension of the prairie-chicken's range exposed it to progressively more arduous winter conditions. The use of snow burrows for overnight roosting during periods of heavy snow has been reported in places such as North Dakota and Wisconsin, and is fairly common in all grouse when they are exposed to deep, soft snow, an excellent insulating medium. At least in the Dakotas, Iowa, Minnesota, and Wisconsin, significant seasonal shifts of the birds toward heavier or more food-rich cover during winter was also known to occur, and similar movements may have occurred elsewhere as well. The birds, especially females, sometimes moved to nearby areas of heavier wooded cover, where elevated foods such as acorns and perhaps green buds might be available. Males seem less prone to move far from their breeding grounds, perhaps because of their territorial tendencies.

Substantial movements to less snowy regions to the south also occurred, and at such times the birds often engaged in surprisingly long-distance flights while en route. In the 1930s, Arthur C. Bent mentioned a flock of 300 prairie-chickens that was seen flying up the shoreline of Green Bay, Wisconsin, apparently about to undertake a flight to the Michigan side of the bay, some 12 to 15 miles away. Similarly, at least two prairie-chickens were observed to arrive during April on an island in the middle of that same bay, about 7 miles from the nearest shore. A few weeks later the birds were seen flying east toward the Wisconsin mainland. Comparable long-distance flights have been reported and probably still regularly occur in high-arctic ptarmigans as they move from open tundra to timberline habitat with approaching winter.

The prairie-chickens arrived in the Red River Valley of North Dakota about 1880, just as that flat and fertile valley was being opened to homesteading and was gradually being transformed from tallgrass prairie to small-grain farming. By the early 1900s the birds could be found in abundance throughout the entire state, with the exception of the far-western badlands. According to Morris Johnson they were sold at local North Dakota markets during the 1880s at about twenty-five cents each, but if they reached Minneapolis markets they brought sixty cents each. Meat markets of Chicago were selling about half a million prairie grouse per year during the early 1870s.

Even though a law passed by the Dakota territorial legislature in 1877 restricted the shipping of game outside the Dakota Territory, birds continued to be sent by train from Fargo to Minneapolis and Chicago well into the 1900s. During this time uncounted numbers of birds were also shot for sport and simply left to rot where they fell. Many others were trapped for personal consumption or for sale; in 1896 several thousand grouse from North Dakota were confiscated by authorities in Saint Paul, the birds having been illegally trapped in the Williston area and shipped by train by people using aliases to avoid prosecution.

As with birds farther north, the prairie-chickens in North Dakota flourished for only a few decades. The birds initially became so common as to be regarded as near-vermin by farmers, in whose fields they feasted. The birds quickly declined and disappeared from most of the state by the early 1930s. My mother, who grew up on a homesteaded farm in the Red River Valley near Fargo in the early 1900s, often heard or flushed prairie-chickens as a youngster when riding horseback through the family's hay meadows. Growing up 30 years later in that same general region, I never saw a live prairie-chicken during my entire childhood. Nevertheless a few birds persisted in a sandy delta of the Sheyenne River near my mother's home, an area of relict prairies that would eventually become a central nucleus of the Sheyenne National Grasslands.

North Dakota became a mecca for wealthy eastern sport hunters before the end of the 19th century. Improved ammunition, the introduction of repeating and automatic shotguns, and increased mobility via automobiles and trains brought great numbers of East Coast sportsmen out into the Dakota prairies. Railroad magnates brought groups of friends out for railroad car hunts, complete with high-quality hunting dogs, fine food, and all the luxuries of a modern African safari. Railroad workers often joined in the fun during work breaks. One contemporary account described a pile of approximately 300 prairie-chicken carcasses that had been shot and abandoned by railroad workers in Dickey County around 1890.

During these decades of great game abundance in North Dakota, there were few restrictions on sport hunting. Game laws existed before 1900, but little if any attention was paid to them. No limit on the shooting hours or number of grouse killed was established throughout Dakota Territory until statehood was achieved in 1887, when a daily limit of 25 was imposed. But by 1915 public attitudes were changing, and a daily limit of 5 grouse per day (of either sharp-tails or prairie-chickens) was established. The season opening was also delayed until mid-September to reduce hunting pressures on very young birds. By the late 1930s and early 1940s the estimated North Dakota kill of prairie-chickens was still around 30,000 to 45,000 per year but was gradually declining, and the last open season on the birds was held in 1945.

In the last decades of the 20th century nearly all North Dakota's greater prairie-chickens were concentrated in two areas. These included the Sheyenne National Grasslands in Richland County and the Prairie Chicken Wildlife Management Area in northern Grand Forks County. There was also a small group of birds using a single lek in southwestern Sargent County near the South Dakota line. The biggest of these populations was in the Sheyenne National Grasslands, where spring lek counts have been conducted since 1963. This population seemingly peaked at about 400 males in the early 1980s and has been in a gradual decline since then. The 1997 census revealed fewer than 100 prairie-chickens but a much larger number of sharp-tailed grouse. Interactions with sharp-tailed grouse and high levels of nest predation have been suggested as possible reasons for this declining prairie-chicken population. The Prairie Chicken Wildlife Management Area near Grand Forks was first surveyed in 1954 when 11 males were seen. Annual surveys began in 1965, but by 1980 the prairie-chickens had disappeared. A restocking of birds began in 1992 and continued through at least 1998, with more than 50 males seen in 1997. Floods in the springs of 2000 and 2001 were expected to have serious effects on the Grand Forks population. It would seem that the total North Dakota population is now less than 1,000 birds.

A similar story for Minnesota was summarized by Max Partch. Prairie-chickens first reached the southeastern part of the state in the 1830s, moving northward and westward at the impressive rate of about 10 miles a year and reaching the Manitoba border some 50 years later. Logging in the central and northern parts of the state opened these regions to grassland and farming. In areas too sandy for agriculture a second-growth brush and forest-edge community developed, which is more favorable to sharp-tailed grouse than to prairie-chickens. By the mid-1800s both species were abundant in central and north-central Minnesota, but around the turn of the century the prairie-chicken began its downward trend.

As early as 1916 a prairie-chicken sanctuary of nearly 150 square miles was established in Polk County by sportsmen from the Crookston area. At about the same time that the forest clearings were gradually being overgrown and the birds were declining, a closed season was initiated in the national forests of Minnesota. From 1919 to 1933 the prairie-chicken season was opened only on an alternate-year basis, with ruffed grouse being hunted in the intervening years. Then, in 1936 there was a closed season for four years. The last open season in the state was held in 1942 when an estimated 58,300 birds were killed. The peak annual kill from 1921 onward occurred in 1923, when more than 400,000 birds were taken.

The end of legal hunting did little to stop the downward population trend in Minnesota's prairie-chickens. The last breeding records for several south-central counties occurred from the 1930s to the 1960s. There the birds disappeared from four counties at the rate of about one county per decade, from Hennepin County in 1929 to Morrison County in 1965. In the mid-1950s a lone male near Hibbing reprised the story of the last male heath hen. It displayed alone on the runway grounds of the Hibbing Airport during two springs, sometimes even attacking incoming planes. Eventually it met its inevitable and quixotic fate, learning too late that tilting at propellers is far more dangerous than tilting at windmills.

In 1973 the Minnesota Prairie Chicken Society was formed to educate the public, promote resource management, and provide research guidance for conserving prairie-chickens in the state. Largely through the efforts of the Nature Conservancy, the following year brought the establishment of the Minnesota Prairie Chicken Preserve System. One important function of this system was to coordinate annual booming ground surveys by volunteers and conservation agency personnel. Minnesota's prairie-chickens have been carefully monitored annually since 1974. With some oscillations, the number of booming grounds has shown a gradual upward trend, ranging from as low as 60 to as many as 140. The total number of males present have ranged from about 500 to more than 1,800, and in 1999 about 1,400. Thus the state spring population should be at least 3,000, and the fall population 5,000 to 6,000, By the late 1990s prairie-chickens were mainly located along the eastern edges of the Red River Valley in areas of glacial moraines and outwash.

The greater prairie-chicken occurred historically throughout the prairie grasslands of the southern half of Wisconsin north to about Saint Croix County. Evidently in Wisconsin it underwent the same kind of rather short-lived population and northward range expansion that occurred in Minnesota following the development of small-grain agriculture and logging. By the 1840s prairie-chickens were selling for as little as twenty-five cents a pair in Milwaukee markets, and tens of thousands were being shipped to Chicago by train. The population explosion and expansion northward lasted a few decades. Although prairie-chickens expanded their range to every northern Wisconsin county by 1920, they were already well past their peak abundance in the state. By 1885 the birds' original prairie range in the southern half of Wisconsin was on the decline even though new habitats were still opening to the north through logging.

As forests began to regrow, the golden period of the late 1800s was followed in the early 1900s by a fragmentation of the species' range. In the south the amount of land in crops soon outstripped that in native grass, and improved fire control in the north allowed for the regeneration of forests. The prairie-chicken's range began to fragment and shrink toward the middle of Wisconsin, gradually retreating to a core central area, primarily in Portage, Wood, Waushara, and Adams counties. In those counties a natural refuge was provided by a combination of lands poorly adapted for farming, such as too-sandy soils, peat marshes, and poorly drained clay bottomlands. Wisconsin was one of the first midwestern states to limit the open hunting season on prairie-chickens. It restricted the season to about three months in the 1800s. During the early 1900s further hunting restrictions were imposed, and in 1917 a multi-year moratorium on hunting prairie-chickens was initiated. After that, only periodic hunting was allowed up until 1956, when hunting was permanently stopped.

As early as 1929 long-term research on prairie-chickens was pioneered in Wisconsin by Alfred O. Gross, who was already famous for his studies on the heath hen. During the late 1920s Aldo Leopold also began his own research on prairie-chickens, and in 1930 he and a student, F. G. Schmidt, produced the state's first population estimate, of about 55,000 birds. The work of Leopold, and especially that of his students and their own academic progeny, has continued with few interruptions more or less to the present time. Important studies were begun by Leopold's students Fred and Frances Hamerstrom in the mid-1930s and continued by them and others on into the early 1970s. Much of this work has been done on the 50,000-acre pseudoprairie called Buena Vista Marsh in Portage County, where an unsuccessful drainage project left a legacy of failed farms and marshy grasses. Aldo Leopold estimated that 4,000 prairie-chickens existed in all Portage County in 1930; by 1950 the population of males at Buena Vista Marsh, the county's major grouse nucleus, was reduced to about 600.

In 1958 the nonprofit Prairie Chicken Foundation was established in Wisconsin for the purpose of purchasing prairie-chicken habitat with private funds, and two years later the Milwaukee-based Society of Tympanuchus Cupido Pinnatus was also formed with the same goal. Through these organizations and state agencies a nucleus of protected lands was acquired, the most important of which were the Buena Vista Marsh in southwestern Portage County, the Paul Olsen Prairie Chicken Management area in southwestern Portage County and adjacent Wood County, and the Leola (Plainfield) Area in Adams and adjacent Waushara counties. By 1980 nearly 11,000 acres had been acquired in the Buena Vista Marsh area, about 1,800 acres in the Paul Olson Area, and 900 acres in the Leola Area. Nearby, in south-central Marathon County, the 27,000-acre publicly owned Mead Wildlife Area is used by prairie-chickens, although it is managed mainly for waterfowl.

These areas became the centerpiece locations for the most complete studies on prairie-chicken biology ever done, largely under the impetus and direction of Fred and Fran Hamerstrom with a host of hard-working volunteers who were fondly termed "Gaboons." Over a 22-year period, more than 6,000 blind-mornings were spent observing lek behavior: documenting lek sizes, seasonality of display activities, and differential individual mating success. Nearly 1,900 prairie-chickens were leg-banded for studies on longevity and movements, and more than 150 of these were wing-tagged for ready field identification. At least 30 publications have resulted from this research, which still provides the basic nucleus of information on prairie-chicken biology and management.

Michigan underwent a somewhat similar pattern to Wisconsin in its prairie-chicken populations, but with worse consequences. Prairie-chickens, once abundant on the prairies and oak savannas of southern Michigan, began to increase in the middle 1800s with forest removal and conversion to farms and hay lands. Yet as they expanded in the north, they declined in the south, with major declines in southern Michigan between 1860 and 1870. By 1875 the birds were thriving in the northern part of the state but were largely limited to marshland areas in southern Michigan. They were widespread in the once-forested northern Lower Peninsula from the 1920s to the 1940s, but by the 1950s the species was declining even there. By 1968 the total state prairie-chicken population was about 200 birds, and by 1980 only 20. The species was completely extirpated from Michigan by 1983.

As of the year 2001, Illinois had managed to escape Michigan's fate, but the outcome is by no means certain. At one time Illinois prairies covered about 8.5 million acres, or nearly two-thirds of the state. Although numerical estimates are lacking, historical accounts indicate that prairie-chickens were common during the mid- to late 1800s. Illinois probably underwent an early explosion of prairie-chicken numbers in the early 1800s as the Illinois prairies were being converted to small-grain agriculture, but documentation is lacking. Judging from 1906 surveys in central and northern Illinois that were later summarized by James Herkert, the species comprised less than 1 percent of the prairie bird population by the early 1900s. In 1933 the state's prairie-chicken population was estimated at 25,000 birds. By the 1950s prairie-chickens were gone from the entire northern region, and they were listed as endangered in the state by the 1980s.

Research on the Illinois population of prairie-chickens and efforts to prevent its extirpation began in the 1960s. Land acquisition and habitat management have been directed primarily toward saving and developing grassland habitats in Jasper and Marion counties in the Prairie Ridge State Natural Area. These efforts were initially begun through the efforts of the Illinois Department of Natural Resources, the Illinois chapter of the Nature Conservancy, and a private nonprofit group, the Prairie Chicken Foundation of Illinois. By 1998 there were about 2,400 sanctuary acres in the two counties. These early conservation efforts of the 1960s and early 1970s seemed to be helping, with a 400 percent population increase in prairie-chickens during those dozen or so years. However, during the middle 1970s the Illinois population plummeted, perhaps owing to a combination of land-use changes, increased nest predation rates, and unfavorable interactions with ring-necked pheasants. By the late 1980s and early 1990s the situation was critical; the few remaining populations were small, fragmented, and in increasing danger of genetic degradation through inbreeding. A low point was reached in 1994 when only 46 birds were estimated to occur in Jasper and Marion counties, as compared with more than 200 birds there in the early 1970s, and about 2,000 birds scattered across 16 counties in 1962. All the populations on unmanaged lands in the state had disappeared during that same 32-year period.

In the early 1990s a decision was made to provide Illinois's remaining prairie-chickens with an infusion of new genetic stock, and 518 birds were imported from Minnesota, Kansas, and Nebraska. By the spring of 1998 the Jasper County flock had rebounded to 84 males on 10 booming grounds, and the Marion flock to 40 males. Nearly all the birds in Jasper County were located in only 700 acres of managed grasslands, reflecting a remarkable and perhaps unrealistically high density of about 27 males per square mile. Flock sizes judged to represent minimum viable populations of 100 to 250 males have not yet been attained, and efforts are under way to add additional suitable habitat to the acreage already present.

Iowa probably once supported prairie-chickens across the entire state, except for the extreme northwest. As elsewhere, their population increased with early settlement, peaking in the 1870s or early 1880s when a significant amount of native prairie remained within a mosaic patchwork of farmland and woodlands. According to Mel Moe the birds then started a long, gradual decline. By 1950 they were limited to Ringgold, Wayne, and Appanoose counties, all located on the Missouri border. The last known nesting in the state occurred in 1952, in Appanoose County.

Attempts to reintroduce the birds into Iowa began in 1982, with 100 prairie chickens from Kansas released near Onawa, in the loess hills of the Missouri Valley. This effort evidently failed. In 1987 another 254 birds were taken from Kansas and released in the Ringgold Wildlife Area. At least some of these birds moved across the Missouri line to establish a booming ground on a private cattle ranch, at a site that had supported a booming ground in the 1960s. Later releases were made elsewhere in Ringgold County, as well as at a site about 55 miles northwest near Orient, in Adair County. Efforts by the Missouri Conservation Department have supplemented these birds in the Missouri–Iowa border region, with booming grounds being newly established in Iowa's Decatur County and in Missouri's Sullivan and Harrison counties. By spring of 2000, 44 males were known to occur on six booming grounds. Iowa's population of greater prairie-chickens was clearly only marginal at the start of the new millennium; its depressing history has been recounted by James Dinsmore.

The story in Missouri is by now all too familiar. The state was once probably made up of about 27–40 percent native prairies. It is likely that prairie-chickens occurred wherever there was prairie, and even in such well-wooded regions as the Ozarks. The birds' peak population, which probably occurred in the 1860s, might easily have numbered in the millions. The last hunting season was held in 1907. The earliest prairie-chicken survey, done in the late 1930s, suggested there were about 27,000 square miles of prairie-chicken range and a population of perhaps 90,000–180,00 birds. A more thorough and realistic study by Charles Schwartz in the 1940s produced an estimate of only 2,500 square miles of occupied range, probably representing about 13 percent of the state's original prairie habitat. At a population density of 3 to 6 birds per square mile, at least 10,000 birds might have been present. Schwartz himself suggested a state population of nearly 14,000 total birds (both sexes) in 1942, but reduced the estimate to 9,250 by 1944. These figures translate into an estimate of about 8,100 males in 1942 and 6,800 males in 1944. By 1983, following new surveys, the prairie-chicken estimate for the state was further reduced to 632 square miles of occupied range and a population of about 5,700 males. A 1988 estimate suggested a total of 3,000 birds. In 1998 Larry Mechlin and others estimated that there were then only 400 to 500 square miles of occupied habitat, supporting only about 1,000 birds. Most of the state's population is now concentrated in three small areas of the Osage Plains in southwestern Missouri. One area, adjacent to the Kansas line and continuous with that state's population, is in western Barton County. Another is in southeastern Barton County and adjacent parts of Jasper Dade, and Lawrence counties. The third major population is in southern Pettis and adjacent northern Benton County. These populations have all been declining since the late 1960s. There are also small relict populations in Audrain and Carroll counties, both north of the Missouri River. In the spring of 2000 an attempt at a complete booming ground survey resulted in a count of 252 males, down about a third from each of the two previous years. In 2001 the count revealed 281 males, but numbers were down in five of nine areas.

Birds from Kansas and Nebraska were released in the 1990s by Missouri Department of Conservation biologists in Sullivan County with some success, and others were released in Mercer, Putnam, and other northern counties. A high rate of conversion of marginal farmlands to Conservation Reserve Program (CRP) lands has improved the prospects for these efforts, especially in Harrison County. In 1966 the Missouri Prairie Foundation was formed with the purpose of acquiring and preserving native prairies, and its efforts have greatly helped the prospects for saving at least a remnant flock of prairie-chickens in the state. The prairie-chicken is listed as endangered in Missouri.

South Dakota is the first state discussed here to have retained a huntable population of greater prairie-chickens. The same is true of Nebraska and Kansas, and Colorado and Oklahoma have also had recent if limited seasons. As in North Dakota, prairie-chickens probably followed the plow west across the state beginning in the 1870s and in turn were followed by market hunters. The birds probably never reached the state's extreme western limits with its dry shortgrass prairies and probable increased competition with sharp-tailed grouse. By the end of the 19th century prairie-chickens were declining in eastern South Dakota and were gone from some areas as early as the 1920s. Wintering birds from North Dakota may have somewhat obscured the overall South Dakota distribution, especially in the north.

By 1968 the South Dakota birds were largely concentrated in and on both sides of the Missouri Valley, especially eastward on the glaciated uplands. That year the first total state estimate of 80,000 birds was made. By 1982 this estimate was reduced to about 39,000 birds, and the overall range was judged at slightly more than 9,000 square miles, representing a density of 4.3 birds per square mile. These numbers might seem rather conservative, especially in view of the fact that estimated hunter kills have ranged as high as 174,300 birds for both species of prairie grouse. Of the two species, prairie-chickens comprised anywhere from 10 to 25 percent of the annual total grouse kill, which in 1994–1999 averaged about 80,000 birds. Thus the probable prairie-chicken kill averaged about 8,000 to 20,000 birds during the late 1990s. State biologists have judged that this number represents 15–30 percent of the total fall populations, which therefore might be in the vicinity of 25,000 to 50,000 prairie-chickens. Ronald Westemeier and Sharon Gough have suggested a "most liberal" 1997 estimate of 65,000 total birds for South Dakota. If this number is anywhere close to accurate, it would mean that South Dakota probably has the third-largest remaining population of greater prairie-chickens, after Kansas and Nebraska.

Lek counts in South Dakota began as early as 1956 and have continued annually ever since. The 42-year average (1956–1997) is 7.02 males per lek, and 0.67 males per square mile. Converting these figures to the 9,000-square-mile range, the both-sex spring population would be about 12,000 birds, and the fall population about double that, or close to the 25,000-bird estimate based on hunter kills. No clear long-term trend is evident from these figures, and the populations have tended to oscillate in synchrony with those of sharp-tailed grouse. Not surprisingly, their populations have declined in drought years and increased during years of cool, moist summers.

Nebraska's prairie-chickens, like those of South Dakota's, occur in a region where tallgrass prairies blend with midgrass and shortgrass habitats, and where sharp-tailed grouse progressively wrest dominance from prairie-chickens in these drier grasslands. During the latter half of the 1800s the prairie-chickens moved west with settlement, gradually overlapping the range of the sharptails. After the passage of the Free Homestead Act of 1863, settlements traced the river valleys westward, along the Platte, the Niobrara, and smaller rivers of the state. The railroads soon followed, providing a means of shipping wild game to the cities farther east. As a result, market hunting became an important means of livelihood, and in 1874 an estimated 300,000 prairie-chickens were shipped out of 30 counties in eastern and southeastern Nebraska. According to one account, 20,000 birds were shot or trapped in Pawnee County alone during 1874, and a similar number were shipped out of Lincoln the following year to eastern markets. At that time the birds were bringing about $4 per dozen in

Chicago. The first game laws of 1877 limited the kill to 75 birds per day and stopped spring hunting.

The market hunting economy dried up in the early 1900s as legal restrictions became more effective. Grouse populations remained high until the 1920s, when they began to decline and fragment. The largest fragment survived along the eastern edge of the Nebraska Sandhills, where sandy soils helped maintain most of the countryside in native grasses. There small-grain agriculture could be undertaken in the firmer soils, especially if supplemented by irrigation. This 20,000-square-mile region was opened to farming development by the passage of the 1904 Kincaid Act and was almost as quickly colonized by prairie-chickens. Both farmer and prairie-chicken almost went bust during the financial crisis of the late 1920s, when farms in the Sandhills began to fail and the percent of land still remaining in natural grasslands over the rest of the state became too small to support prairie-chickens.

Hunting of the birds was terminated statewide in 1929, and soon thereafter the dust bowl descended on Nebraska and the other plains states, bringing grouse populations to an all-time low in the late 1930s. The season was reopened in 1950 and has been allowed nearly every year since then. There are no good estimates of grouse populations for these early years, but annual surveys of leks and hunter kills, as well as other survey observations, have been performed since 1955, and yearly statewide estimates have been available since 1986. In contrast to Kansas, which releases detailed descriptions of how its population and hunter-kill data are obtained and their statistical reliability, the Nebraska Game and Parks Commission provides no such supporting information. As a result, the accuracy of the commission's figures is unknown. The estimated state's population has ranged from about 220,000 birds in 1987 to 102,000 in 1992.

A gradual population decline in Nebraska's prairie-chickens occurred during the late 1980s, but the start of the CRP, a program encouraging farmers to plant cover plants such as grasses on marginal farmlands, has probably been responsible for an apparent population upswing in the 1990s. Using the estimates just mentioned, one might judge that the Nebraska population of prairie chickens during the late 1990s was about 100,000 birds. Between 1986 and 1996 the average statewide total grouse kill was estimated at 73,650 birds, of which some 44 percent, or slightly more than 32,000, were prairie-chickens. This would mean about one-third of the population was then being killed annually, easily the highest harvest rate for any state where prairie-chickens are legal game, both in numbers of birds shot and in percentage of the population being removed by hunting. Ronald Westemeier and Sharon Gough's 1997 estimate was 131,000 total birds for Nebraska's sandhills region, which holds the vast majority of the state's birds. The estimated annual hunter kills of prairie grouse during the 1998 and 1999 seasons averaged about 47,000 birds. With prairie-chickens historically comprising approximately 45 percent of the total kill, about 21,000 birds might be a reasonable estimate of annual Nebraska prairie-chicken kills during the late 1990s.

In the grassy high plains of eastern Colorado, the sharp-tailed grouse has long held sway as the grouse of the steppe-like grasslands. Not until the late 1890s did the greater prairie-chicken first appear in the state, the birds probably having followed the South Platte Valley west into Yuma County. Eventually they found their way into at least 7, and possibly as many as 11, Colorado counties, extending at their peak abundance from Larimer County in the northwest to Kiowa County in the southeast. But, as elsewhere, the drought years of the 1930s caused a rapid decline, and the population eventually retreated to the South Platte valley of Yuma County. The hunting season in Colorado was closed for the first time in 1929, and essentially terminated (at least until 2000) in 1937.

The earliest statewide estimate of Colorado prairie-chicken populations was made in the early 1950s, when 2,000–3,000 birds were thought to be present, based on winter counts and lek counts. A second estimate, 700–800 birds in 1963, was based on limited survey data but did suggest that the population had seriously declined during the previous decade. By that time the birds were mostly concentrated in Yuma County, with smaller populations in Logan, Washington, and Phillips counties. In 1973, with a statewide estimate of only 600 birds, the greater prairie-chicken was declared an endangered species in Colorado.

During the 1980s and 1990s significant habitat management and reintroduction efforts were initiated, with the result that the core Yuma County population expanded and new populations were established in Logan and Washington counties. By the late 1990s the birds were believed to be relatively secure in Yuma, Washington, and Phillips counties, and reintroduced populations were present in Weld, Morgan, Logan, and Sedgwick counties. In 1998 the species was removed from the endangered category. At that time the state's greater prairie-chicken population was believed to number 8,000–12,000 birds, and a highly restricted, permit-only hunting season was initiated in Yuma County.

Eastern Kansas has long represented the heart of the interior greater prairie-chicken's preferred range. From the state's northern border in Marshall County to its southern border in Cowley and Chatauqua counties, a variably wide green belt of bluestem prairies once grew in profusion over shallow, calcareous soils rich in flint and chert. These rocky soils of the Flint Hills made extensive cultivation almost impossible except on the adjacent black-soil lowlands and even today have prevented east-central Kansas from becoming a wall-to-wall cornfield. To the east of the bluestem prairies lies a transition area of native grasses and cropland, where there is an approximate three-to-one ratio of native grasses to feed crops and other cover. This ratio provides a near-perfect combination of summer nesting and brooding grassland habitat, complementing nearby fall and winter food and escape cover, and resulting in the optimum prairie-chicken habitat in the state. Somewhat farther south are the blackjack oak prairie savannas of southeastern Kansas, where sandy soils and wooded uplands are continuous with the cross-timbers region of Oklahoma. Here, a savanna-like community of large oaks, growing singly or in groves, is interspersed with prairie grasses, and these areas too are locally used by prairie-chickens.

The prairie-chickens of Kansas are among the best-studied population of prairie grouse anywhere. After baseline studies by Maurice Baker in the 1950s, the state's Department of Wildlife and Parks biologists have undertaken exhaustive field studies and population surveys, especially on the greater prairie-chicken. Much of this work was analyzed in the 1980s by Gerald Horak and has been summarized by Horak and Roger Applegate.

Judging from booming ground surveys, the range of the greater prairie-chicken has declined considerably in Kansas since the 1960s. From a density of about 12 birds per square mile in the 1960s, there was a drop to about half that in the early 1970s, followed by a high point of 11.1 birds per square mile (4.38 per sq km) in 1980. Since then there has been a rather consistent decline, down to about 6 birds per square mile (less than 2 per sq km) in the late 1990s. Comparing the overall 1963–1989 average density of males against that of the 1990s, there has been a 27 percent decline. A total of 178,000 birds was Ronald Westemeier and Sharon Gough's "most liberal" 1997 estimate, compared with 1 million birds in 1989. Kelly Cartright was unable to correlate these population declines with specific land-use changes during recent decades, such as changes in grazing intensity or timing and burning regimens. Surveys in 2001 indicated no significant rangewide changes from the previous year.

Hunter surveys provide another source of judging population trends in Kansas, and these data are available starting in 1957. Their results suggest an overall stable harvest rate of about 45,000 birds annually. However, since the 1990s the harvest has dipped well below 20,000 birds for the first time since the 1960s. By the spring of 2000 the Kansas population index for greater prairie-chickens was down 6 percent from the previous year, with the eastern and southeastern populations declining most, and the western ones increasing as a result of CRP grassland conversion. Data obtained from rural mail carrier surveys in Kansas are similar but are more variable and probably less reliable. In 1980 the Kansas greater prairie-chicken population was estimated at 200,000 birds, which probably represented a high point for the previous 40 years, judging from both booming ground and mail survey data. During the next twenty years these two indexes declined to about 30 percent of their 1980 highs, putting the Kansas population in the vicinity of 60,000 birds. The 1998–1999 estimated harvests were of about 12,000 birds, or around 20 percent of this total. Both these figures fall below Nebraska estimates in the same time frame.

Oklahoma represents the current southernmost range of the interior greater prairie-chicken. It certainly once occupied essentially all the eastern two-thirds of the state, except for the highly wooded areas of the southeast. As elsewhere, it had a short-lived flush in the early 1900s associated with agriculture, but this period soon ended and a period of slow decline set in. No early estimates of the Oklahoma population were made, but by 1943 the birds' range had been reduced to 4,065 square miles, with a population of about 12,600 birds. By 1979 this range had further diminished to 2,355 square miles, and its state population was judged to be about 8,400 birds. The majority of these were concentrated in Craig, Mayes, Rogers, and Nowata counties. The rest consisted of isolated populations in Payne, Tulsa, and Ottawa counties. The average population density for the entire Oklahoma range was judged to be less than four birds per square mile.

As of the late 1990s Oklahoma's population of greater prairie-chickens had been further reduced, the rate of decline increasing since 1990. The area of occupation then included one major population in northern Osage County and extreme northeastern Kay County. Part of this population occurred within the present-day Tallgrass Prairie Preserve, and these birds represented the southern limits of Kansas's Flint Hills and Blackjack flocks. The second population fragment included the northern parts of Nowata and Craig counties, plus extreme northwestern Washington County. An isolated flock still existed in central Noble and adjacent western Pawnee County; another was in the area where Rogers, Wagoner, and Hayes counties meet; and the last small flock was present in Ottawa County, northwest of Miami. Payne County may also have supported some birds.

The 1979 population estimate of 8,400 birds for Oklahoma does not seem to have been officially updated. This number may be unreliable, since hunter harvest that year was 12,500 birds, a figure that included some lesser prairie-chickens. Between 1982 and 1998 the overall density index (based on booming ground counts) for the state dropped from more than 8.0 to less than 2.0, a reduction of about 80 percent. Ronald Westemeier and Sharon Gough's 1997 estimate for Oklahoma was a total of 1,500 birds. Lek surveys for 1999 and 2000 showed further declines. By the spring of 2000 an average of only 3.3 males were present per historical lek site, and the lek density was only 0.18 leks per square mile. If these statistics provide a realistic basis for estimating the state's greater prairie-chicken population, it may have been in the vicinity of 1,000 birds, an all-time low, although Oklahoma biologists have challenged this number, suggesting that 8,000 to 10,000 may be closer to the truth. The last legal hunting season in Oklahoma was held in 1997, which was a two-day season with a season limit of two birds.

The states that still hold relatively secure populations of at least 1,000 interior greater prairie-chickens include Nebraska with possibly as many as 100,000, Kansas with 60,000, South Dakota with 25,000–50,000, Colorado with 8,000–12,000, Oklahoma with 1,000–10,000, Minnesota with 5,000–6,000, and Missouri with 1,000. Whereas all these estimates involve substantial uncertainties if not outright guesswork, they provide at least a collective ballpark figure of about 200,000 to 250,000 birds as of the late 1990s. This compares with the million I estimated to exist in the early 1970s, and a half million in the early 1980s. If these figures are at all realistic, the species has declined by 80 percent in the last three decades, a similar rate of decline to that which has occurred in the lesser and Attwater's prairie-chickens.

GENERAL REPRODUCTIVE BEHAVIOR AND TERRITORIALITY

More work has been done on the reproductive behavior of the interior greater prairie-chicken than the research directed toward all the other prairie grouse combined. Ironically, this is partly the result of the birds being sufficiently rare in most states to warrant special attention from private, state, and federal conservation agencies, whereas the much more widespread and abundant sharp-tailed grouse has been relatively neglected over most of its range. There is also the "glamour" effect of working on such a magnificent bird; far more research has likewise been done on the American bald eagle than the more widespread golden eagle, in large part because of the symbolic influence of the former. Two states in particular, Kansas and Wisconsin, have devoted an unusual degree of research attention to the interior greater prairie-chicken. Kansas has done so largely because this species has long been the state's most important upland game bird. Wisconsin has been active not only because of the species' relative rarity there but also as a result of its devoted following by an unbroken succession of ecologists and ornithologists whose academic and intellectual genealogies can be traced back directly to Aldo Leopold. Owing to the extent of comparable data for these two states, they provide a convenient central focus for discussion and comparison.

During the winter, prairie-chickens assemble in loose flocks, or "packs," those in the northern parts of the range probably moving farther from booming grounds and nesting areas and into heavier cover than those farther south. Wherever food sources and snow conditions permit, the males remain quite close to their leks. Thus Maurice Baker observed some midwinter activity of males on a lek during a mild Kansas winter, and the Hamerstroms noted that the birds sometimes begin display activities in Wisconsin while there is still snow on the ground. No banded adult males in Wisconsin were ever recaptured more than nine miles from their associated lek, and 75 percent of 588 banded adult males were recaptured within three miles of their lek.

In the Buena Vista Marsh area of Portage County, Wisconsin, a total of 48 leks existed, 38 of which were "stable" leks that persisted from year to year; 10 leks were of uncertain permanence status. A maximum of 550 males were present in 1950, the first year of the study and the year of peak male attendance. The study area encompassed about 70 square miles, so the lek density approximated a lek per 1.5 square miles, and the male density was about 6 males per square mile. By 1971 the population had dropped to 198 attending males on 21 leks, and the mean number of males per lek declined from 13.5 to 10.3 on the stable leks. The 22-year average from 1950 to 1971 was 9.0 males per stable lek. The largest number of attending males observed at any stable lek was 45; the smallest was a single bird.

During the Hamerstroms' Wisconsin study, the number of active leks tended to increase with increasing overall male populations, as did the number of attending males per lek. By early April, the number of males present daily on any single lek tended to become more stable, reaching a peak in the third or fourth week of April. There was then a drop in male numbers, followed by a secondary peak in mid-May when females were probably being fertilized for renesting. The pattern of female attendance on the leks followed a fairly predictable pattern, starting in early April, peaking between April 14 and 22, and then dropping off. The earliest copulation was seen on April 6. Most occurred between April 18 and 28, a few days (average, 3.2) after the maximum rate of visitation by hens. This would suggest that not all females are mated successfully the first day they visit the lek. A second, smaller peak of matings occurred between May 7 and 13, presumably reflecting renesting efforts,

Of the total copulations seen, 416 were by known-age males. Of these, 82 percent were by birds more than a year old, and 18 percent by first-year males. Of 483 copulations by older known-age males, 74 percent were performed by those with interiorly situated territories. However, of 72 copulations performed by first-year males, two-thirds were by males holding peripheral territories. Clearly there is a great reproductive advantage to those males able to obtain and defend interior territories, and the majority of such territories are held by older birds. However, some young males holding peripheral territories did occasionally manage to mate successfully, achieving 48 of 483 observed copulations (10 percent). Although this is a small percentage, the possibility, however slight, of young, relatively inexperienced males passing on their genes probably makes lek attendance a better gamble for them than displaying solitarily. There did not appear to be any difference in the success rate of attempted copulations relative to successfully completed copulations among older versus first-year males, and the same result applied to these age-groups of females.

The Hamerstroms did not attempt to determine the exact ages of the most reproductively successful males, but they did construct several survivorship tables based on the known postbanding survival lengths of birds banded during winter or on leks over their 20-year study period. There were a total of 762 recaptures of males banded as first-year immatures and recaptured on leks during the spring. Of these, 398 males (52 percent) survived at least to their second winter or spring (thus approaching two years old); 204 (27 percent), to their third year; 104 (14 percent), to their fourth; 40 (5 percent), to their fifth; 9 (1.3 percent), to their sixth; 4 (0.5 percent), to their seventh; 2 (0.2 percent), to their eighth; and 1 (0.1 percent) to the ninth spring following banding. Clearly, the chances of a male interior greater prairie-chicken's living 5 or more years in the wild under these conditions in Wisconsin is probably at most only a few percent—these birds were banded at about 6 to 9 months of age, so any posthatching mortality that occurred during the first summer and fall following hatching, a time of typically high mortality, was unmeasured. The importance of a stable dominance hierarchy, with a smooth "transition of reproductive power" by younger males replacing the older and more experienced males as they die, is clear.

The Hamerstroms found that most copulations occurred on leks having 11–15 males; fewer successful copulations were seen on those with fewer than 6 or more than 26 males. Reduced female attraction to small leks is understandable and perhaps predictable. The reason for reduced copulations on large leks is less obvious, but may relate to the problems a single dominant male might have in effectively maintaining dominance over such a large number of competitors, thus spending more time in aggressive interactions than in direct courtship. The optimal size of leks might also be limited by the inability of a single dominant male to differentiate himself effectively from all other males in the midst of such confusion, thus being less effective in attracting females. It is still not clear just how dominant male prairie-chickens make themselves known to and become recognized by females, but in sharp-tailed grouse they tend to display with greater vigor and persistence than do nondominants. Work by Michael Schroeder has shown that females may visit as many as six different leks before copulating, and the great majority visit more than one. However, a single successful copulation suffices for each nesting attempt. These important female preference traits make the genetic value of being a dominant and unusually sexually attractive male even more powerful, as he might thereby control the reproductive effectiveness not only of the lek he dominates but also of nearby ones.

Near the end of the Hamerstroms' study, Ingemar Hjorth revisited some of these same leks and confirmed many of their findings, including that territories are somewhat flexible in size and shape, and may vary somewhat in the course of a season. Furthermore, their boundaries are often marked by obvious landscape features such as vehicular tracks or ruts. He also determined that the interior leks he and the Hamerstroms studied averaged 221 square yards (185 sq m). Furthermore, although one central male might defend an area as small as 66 square yards (55 sq m), another might occupy a territory twice as large as the mean of all territories on the lek.

In Kansas, Robert Robel and Warren Ballard found that the average size of 6 measured male territories prior to experimental manipulation (selective male removal) was 167 square yards (140 sq m), with the alpha and beta males' territories having a mean of 189 square yards (158 sq m). In another experiment the average of 7 males' territories was 218 square yards (182 sq m), of which those of the alpha and beta males averaged 350 square yards (293 sq m). Among all 13 measured territories the average size was 218 square yards (182 sq m), of which those of the 2 most dominant males averaged 269.7 square yards (225.5 sq m). Somewhat similar studies in Kansas were performed by Gerald Horak. He determined that on one lek, 11 males had territories ranging from 52 to 186 square yards (43.5–155.5 sq m), averaging 102 square yards (85 sq m). The most dominant males on this lek were not identified, but the most interior and presumably most contested territories were smaller than the overall average. Limited information and personal observations would suggest that the larger the number of participating males on a lek, the smaller the average territory size.

Robel and Ballard began manipulating territories and relative male dominance by removing dominant (alpha- or beta-rank) males, those responsible for most of the successful copulations observed on the lek. Of 132 attempted copulations prior to these manipulations, 92 percent had been successful, and 89 percent of these had been by dominant males. After these males were removed, only 13 percent of 39 attempted copulations by the remaining males were successful. These were performed by lower-ranking males that had moved to the center of the lek after removal of the original dominants. Not only was the social organization thus disrupted during the year of male removal, but it carried over to the following spring. At that time there were fewer males on the lek; fewer females visited it; and fewer successful copulations occurred. Once established on a lek, males are likely to return to the same lek in subsequent years and may even defend essentially the same territory. Young males may visit several leks before finally becoming established at one; the maximum number so far recorded being six different leks. Daily attendance on the lek is affected only slightly by wind and temperature. Kelly Cartright found that the single most important factor influencing the number of males present is the period elapsed since sunrise, which is the time of maximum male attendance.

TERRITORIAL AND COURTSHIP DISPLAYS

Each male spends most of the time in his own territory performing a variety of postures, movements, and calls that serve to ward off potential competitors and to attract females. The adult males (Figure 9), as well as their display behaviors, are virtually identical to those of Attwater's prairie-chicken, and so far as is known, the heath hen. The most characteristic of all greater prairie-chickens' displays is the booming posture and call (Figure 10). It is essentially identical in the interior and Attwater's races, and judging from available descriptions was probably identical in the extinct heath hen (compare Figures 2 and 10). In the heath hen it was called "tooting" by Alfred Gross, a term Ingemar Hjorth believed better fit the male's associated vocalization and its associated foot-stamping sounds. He used that term to describe the call proper, whereas he called the postural component the "forward" display.

In assuming the forward posture, the bird's tail is cocked to at least the vertical, the earlike feathers, called pinnae, are variably raised, and the primaries of both wings are lowered while still held within the flank feathers, rather than being spread laterally as in sharp-tailed grouse. The display sequence begins

Figure 9. Interior greater prairie-chicken, male (after photo by the author).

Figure 10. Interior greater prairie-chicken, male booming (after photo by the author). Sonogram after Hjorth (1970); duration 3.0 seconds.

with a rapid foot-stamping that lasts up to two seconds, while the feet are alternately stamped at a rate of about 20 times per second, producing a sound audible up to 100 feet away. The tail is then quickly opened and shut twice, producing a single click audible only at close range, and the first of the three booming or tooting notes are uttered as the yellow air sacs rapidly inflate. The three-noted call varies little in loudness or in fundamental sound frequency, but the third and longest note (lasting about a second) has better developed harmonics, making its average pitch seem somewhat higher. The head is jerked down slightly during the first note, but not nearly so conspicuously as in the lesser prairie-chicken. The air sacs also vary slightly in degree of inflation during the three notes, being most highly expanded during the final note. The call's typical transcription, "Old-Mul-dooon," describes the sound well, a noise much like that produced by blowing over the opening of a large bottle. The tail is progressively spread and again closed toward the end of the vocalization, and the beak finally opens as the air sacs deflate. No special direction is maintained during booming; over time the male is likely to face all directions. Booming is performed by each male at a usual rate of several times per minute, but is especially frequent and intense when females make their appearance on the lek.

Under favorable conditions this call can be heard well over a mile away and may even carry for several miles, making it audible within the home ranges of many other prairie-chickens. In favorable habitats, leks are often spaced little more than a mile apart; under such conditions female prairie-chicken are probably never out of the hearing range of males from at least the nearest lek, and perhaps they can hear calls from several leks. The low-frequency notes are especially well adapted for long-distance transmission over open habitats, in contrast to high-pitched sounds that tend to be easily absorbed by surrounding vegetation and wind. When females are present on the lek, a quite different call is uttered. It is a strange, whooping call, sometimes described as a "poik" note, lasting about half a second (Figure 11). It has a fundamental frequency about twice that of the booming call but is otherwise acoustically similar. There is no obvious associated neck inflation, but at times sharp tail-clicks may precede it, as is the case with booming. It also carries relatively long distances, and this sound source can be more easily localized than can booming.

One specific postural display directed only toward females is also used, and only when the male is close to a specific female and in a precopulatory situation. This is a "bowing" or "prostrate" posture, with the breast lowered to the ground, the wings outstretched to the side, and the pinnae and tail fully cocked (Figure 12). The posture is silent, and may be held for only a few seconds. Often copulation immediately follows. This occurs in the manner of all gallinaceous birds, such as chickens, pheasants, and turkeys, with the female lying with her breast flat on the ground and with her wings spread sufficiently far to provide a stable platform for the male. Copulation is brief, and if successful, the female shakes herself, preens for a time, then leaves the lek rather promptly.

Figure 11. Interior greater prairie-chicken, territorial male calling (after photo by the author).

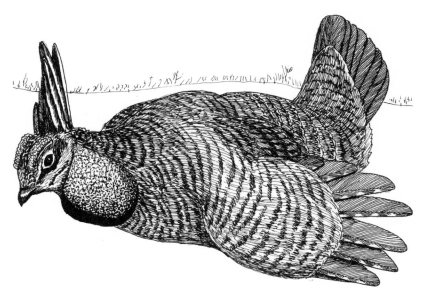

Figure 12. Interior greater prairie-chicken, male in prostrate ("nuptial bow") display posture (after photo by the author).

A moderate percentage of the total copulation attempts are unsuccessful (15–25 percent in Hamerstrom's Wisconsin study), often because of harassment and direct physical interference by nearby males or even by other females.

Short, jumping flights, called "flutter-jumps," are usually initiated when a new bird arrives on the lek, regardless of its sex, but seem especially common and vigorous after the arrival of a female (Figure 13). The flights themselves are silent, except for some wing noises, but immediately after landing the bird utters repeated cackling notes, often a prolonged note followed by several shorter ones. Hjorth termed a similar and even more common call the "staccato cackle." It is usually uttered from an upright posture (Figure 14). It may be uttered when the bird is outsides its territory, such as when its is arriving on the lek or moving through territories held by other males.

Aggressive encounters between males involve several different postures, such as confrontational crouching while facing one another and uttering cackling or whining notes. The birds may also walk parallel to a rival in an upright stance, often simultaneously booming. Fights are most common early in the lekking season while territorial boundaries are still being actively contested. They involve alternate or simultaneous jumping into the air, attempting to strike the opponent with the feet or the wings, or pecking the other bird's throat or breast (Figure 15). Often feathers are pulled out during these contests;

Figure 13. Interior greater prairie-chicken, male flutter-jumping (after photos by the author).

I have seen some bleeding scratches on the bare air sacs, but fatalities have apparently never been reported.

NESTING AND BROODING BIOLOGY

As soon as the female has been successfully fertilized, she leaves the lek and probably begins seeking out a nest site, if one has not already been selected. In several studies the female has been found to place her nest closer to a lek different from the one where she was fertilized. Observations by Daniel Svedarsky indicate that the first egg of the female's clutch is laid about four days after she was fertilized, a somewhat surprising time-lag in view of the simple nest the female constructs. Presumably it takes that long for hormones that might have been triggered by mating to begin egg production, plus the time required for a fertilized egg to pass through the oviduct.

As with other prairie grouse, the female lays an egg a day, occasionally skipping a day, so that a clutch of 12 eggs would take about two weeks to complete. Incubation may begin with the last egg laid, the next-to-last egg, or as late as several days after the laying of the last egg. It is performed only by the female and lasts 23 to 25 days. Nest destruction will result in a second or even multiple nesting attempts. Later clutches tend to be smaller, averaging fewer than 10 rather than 11 to 13 eggs, and the rearing success of any resulting broods is also generally less. Considering 11 separate field studies, the collective average nesting success for interior greater prairie-chicken is about 44 percent, as compared with about 30 percent for the Attwater's prairie-chicken and 28 percent for the lesser prairie-chicken. Repeated efforts at nesting tend to increase overall reproductive success somewhat, and under favorable conditions as many as two-thirds of the females may eventually succeed in rearing broods. The age of the female seems to play no significant role in nesting success, but nests placed in thicker vegetation are more successful, as are those located in areas having relatively few predators. Additionally, higher nest success has been correlated with relatively low litter cover, higher grass and forb cover, and lower woody vegetation immediately around the nest.

The chicks begin to take short flights at about two weeks of age and may later disperse or join other younger birds as they become independent. Young males may visit groups of older males at their autumn leks and by the follow-

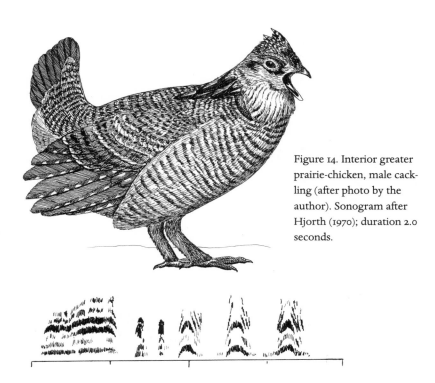

Figure 14. Interior greater prairie-chicken, male cackling (after photo by the author). Sonogram after Hjorth (1970); duration 2.0 seconds.

ing spring are actively competing for territorial positions there. Young birds tend to disperse longer distances than adults, and females of both age groups tend to disperse farther than males of their corresponding ages. In the Hamerstroms' Wisconsin study the majority of recaptured birds of both age-groups and sexes had moved no more than 5 miles from the place of their banding, and only 5 documented movements out of more than 1,000 were of distances greater than 20 miles.

POPULATION TRENDS AND FUTURE OUTLOOK

During the years of burgeoning human populations and increasing agricultural production that followed World War II, the last remaining remnants of tallgrass prairies were quickly converted to cropland. The era of organic and synthetically constructed pesticides that persisted in the environment for long periods before degrading into still other new and potentially dangerous compounds had arrived, and the Endangered Species Act of the 1970s was still only a pipe dream. As of the turn of the millennium, North America had lost about 80 percent of its remaining population of interior greater prairie-chickens in

Figure 15. Interior greater prairie-chicken, territorial fighting by males (after photo by the author).

three decades, while the Attwater's prairie-chicken was being kept on life support by release of captive-raised birds. Meanwhile, the lesser prairie-chicken was in grave danger of dropping into the endangered species category without even the benefit of having first passed through a threatened status.

The greater prairie-chicken has an English vernacular name that sadly understates both its beauty and its aesthetic value. Granted this name makes clear that the bird's presence is a reliable indication of native prairies, and it is somewhat "greater" in size than the lesser prairie-chicken. But the prairie-chickens are no more chickens than a turkey is from Turkey. Perhaps the prairie-chicken should have been called something like "soul-of-the prairie" or "spirit-of-the-grasslands," forcing anybody who wants to kill it to think twice about his motives. Those who have spent a spring sunrise with prairie-chickens will know exactly what is meant by these semantic intimations of the holy; there is a sense of the sublime when one is in the presence of displaying prairie-chickens. They are acting out courtship routines inherited from distant ancestors, on grassland sites used yearly by uncountable generations past. Additionally they are determining, by both battle and bluff, which individual males are most fit to transmit their genes to the next generation by being able to attract females that visit the lek when ready to lay their eggs. Every spring, Darwin's concept of survival and reproduction of the fittest is played out daily on these grassy hilltops. The leks of prairie grouse provide a highly evolved and ritualized mechanism dedicated to attaining efficient and appropriate pairing and mating. Being able to witness these performances may accurately be termed an auspicious act; the actions of the birds provide a reliable augury as to the likely future fortunes of the species. In the sense of having a specific dedicated function, the leks of prairie grouse also represent sacred natural sites.

We have far too few sacred natural sites in the eastern Great Plains; most of the holy sites of the Native Americans that once ruled the plains have been cleared and "developed," or their exact locations have been long forgotten. We must not forget the locations of prairie-chicken leks; they whisper to us of secret places where grama grasses and bluestems grow thick on the ground, and where flint arrowheads are likely to lie buried beneath the thatch and loess. They tell us of meadowlark and dickcissel song-perches and of traditional coyote hunting grounds. They are as much a connection to our past as are the ruts left in the soil by Conestoga wagons or the preserved costumes of Native American cultures carefully stored in museums. But such eroding wagon trails and fading costumes are essentially static and retrospective icons; prairie-chickens are the vital essence of life itself, clinging to their brief moments in the sun with all the energies they can muster. They risk attack by early-rising hawks and by late-flying owls simply to have a chance to reproduce before they are all too quickly cut down by predators, disease, or a hunter's gun. The feathers they wear that are sometimes strewn over the ground when a predator has been successful are the camouflage colors of dead grass, and their soft hypnotic voices are exciting yet soothing, like the mantras emanating from a Hindu temple. They comprise a New World symphony all by themselves, a harmony of sound, color, and movement.

It is easy enough to save these wonderful sights and sounds for following generations. We only need to recognize that both prairies and prairie-chickens need to be preserved, if for no other reason than to help us understand what a plains author such as Willa Cather meant when she wrote lovingly of our "shaggy grass land" or what plant ecologist John Weaver meant when he said that "civilized man is destroying a masterpiece of nature without recording for posterity that which he has destroyed." We may sometimes destroy the things we love out of ignorance; we should never do it purposely.

5

DAWN DANCERS ON DUN GRASS
The Sharp-Tailed Grouse and the Northern Prairies and Shrublands

Watching sharp-tailed grouse and prairie-chickens displaying is something like seeing the same story line as conceived by two different directors and screenwriters. The greater prairie-chicken's display performances are reminiscent of a Shakespearean demitragedy, filled with mournful notes punctuated by occasional wild screams, and a generally dignified, almost statuesque, demeanor by the male participants. The sharp-tail turns essentially the same plot into a comedy, with the males' animated movements marked by pirouettes, short dashes, sudden freezes in motion, and a soft rattling sound rather like an agitated rattlesnake. Yet the endgame is the same: a brief, hurried mating that is often over almost before one realizes it has begun, the avian equivalent of a one-night stand.

Of all the prairie grouse, the sharp-tailed is the only one that has managed to cope with the problems of survival in modern times, largely by the strategy of simply retreating from them. It has done this partly because its historical presettlement range was the largest of all the grouse species that are ecologically adapted at least in part to grasslands, so it has had the greatest opportunities for geographic retreat. In addition, its preferred habitats include a wider diversity of vegetational types than those of any prairie-chickens. Furthermore, among the grasslands that are utilized by sharp-tails, these habitat types are the ones that remain closest to their original ecological state, certainly in overall area if not in actual plant composition. Specifically, they include the short-stature steppe-like grasslands of the western high plains, occurring on lands that because of their limited precipitation are most often used for grazing.

In contrast to the prairie-chickens's preferred habitat, grasslands represent a minority of the sharp-tail's favorite places to live. The species comprises six distinguishable subspecies, each of which has rather distinctive ecological attributes. Before European settlement, sharp-tailed grouse occurred in what are now represented by at least 8 Canadian provinces and 21 states, from Alaska in the north to Quebec in the northeast, and south to California and New Mexico. Three of these races occur only in Canada and Alaska, where their ranges and habitat preferences fall outside the geographic and ecological boundaries of this book. The other three races are geographically organized in a replacement series from west to east in southern Canada and the lower 48 states.

The prairie race *(campestris)* of the sharp-tailed grouse's original western limits generally followed the western limits of the tallgrass prairies from eastern Manitoba south to northern Kansas. From there it extended east to southern Ontario, and south locally to Iowa, Wisconsin, and Michigan. Unlike the grass-loving greater prairie-chicken, however, it is a bird of the tallgrass-forest transition zone, where brushy vegetation might comprise 20–50 percent of the total ground cover. The prairie race is only marginal to our immediate interests, just as it is ecologically marginal to the grasslands themselves. It is also the only one that has expanded its range somewhat, as eastern Canadian forests have been logged, opening suitable habitats for it. Transplants of birds from farther west have aided significantly in this process of eastern expansion.

The plains race *(jamesi)* of the sharp-tailed grouse replaces the Columbian race immediately east of the continental divide, extending from there over the shortgrass high plains eastward to the tallgrass prairies. Its historical northern limits were the southern halves of the prairie provinces to the Manitoba–North Dakota border; from there it reached south through the Great Plains states, from Montana and North Dakota south originally to southeastern Colorado and western Oklahoma or Texas. A small population, now extinct, also occurred in extreme northern New Mexico. Once considered to be part of the plains race, it is now regarded as a separate but vanished race *(hueyi)* that was not even recognized as such until it was already gone.

The Columbian race *(columbianus)* of the sharp-tailed grouse originally occurred from southeastern British Columbia south to northern California, west across the intermountain plateaus to the continental divide in Montana, Wyoming, and Colorado. It is a race especially typical of bunchgrass prairies of the Pacific Northwest and of the sagebrush grasslands of the intermountain regions.

THE PRAIRIE SHARP-TAILED GROUSE

The prairie sharp-tail is somewhat darker and more rufous-toned above than the other two races discussed here, and it has more prominent white spotting.

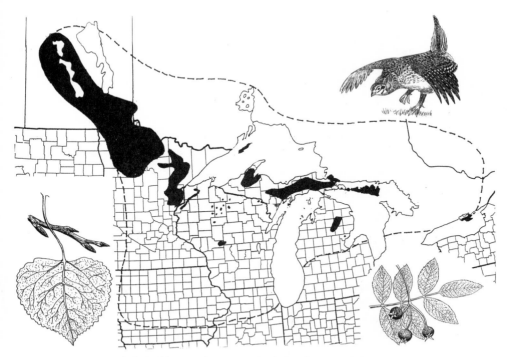

Map 7. Historical (dashed line) and current (inked) distributions of the prairie sharp-tailed grouse. Inset drawings show quaking aspen *(lower left)* and wild rose *(lower right),* important fall and winter food plants.

At its northwestern limits its historical range extended beyond the large glacial lakes in southern Manitoba and into southeastern Saskatchewan, where it merged rather imperceptibly with the plains sharp-tail (Map 7). There it still largely occupies the tallgrass prairie–aspen parkland transition zone. In this region tall grasses fight a continuing battle for dominance with aspen forests, the grasses winning wherever fires are frequent enough to prevent tree establishment; the aspens winning where the fires are suppressed long enough for their saplings to grow out of reach of ground fires. No good estimates of the current combined Saskatchewan–Manitoba population exists, but the prairie sharp-tail probably still occupies something like 50 to 90 percent of its original range in those two provinces, and thus is perhaps the most stable of all the surviving grouse populations.

The Ontario population of prairie sharp-tails is certainly much smaller than those farther west and is mostly confined to Ontario's southwestern corner, where deforestation, fires, and natural brushy openings in the forest provide islands of habitat in a coniferous forest matrix. Its range extends eastward along the northern shores of Lake Huron and includes the large Canadian-owned Manitoulin Island, as well as a small introduced (and possibly now extirpated) population along the northern shore of Lake Ontario. There are also a few islandlike populations of sharp-tailed grouse in the boreal forests to the east and north of Lake Ontario toward James's Bay, but these birds almost certainly belong to or at least are closer to the northern race *phasianellus*. The overall Canadian range of the prairie sharp-tail may be almost as extensive as its original historical one. It still occupies a region with a fairly low human population, at least as compared with prairie sharp-tails occurring south of the Canadian border. Breeding bird surveys for the period 1966–1992 suggest that the Canadian population of the prairie sharptail is increasing over most of its range.

The prairie sharp-tail once extended south to south-central Iowa and northern Illinois, but the birds were gone from Illinois by the early 1900s and were evidently extirpated from Iowa by 1934, although there have been efforts to reintroduce them into the loess hills region along the Missouri Valley near Sioux City. They retreated northward across Minnesota until they occupied only the northern third of the state. They likewise moved out of the oak savannas of southern Wisconsin, where they had once been called the "bur oak grouse," to more coniferous parts of northern Wisconsin and to northern Michigan, especially the relatively forested Upper Peninsula. In all these areas coniferous forests predominate over grasslands, and the prairie sharp-tail now ekes out a rather precarious living in the shadows of forest edges and recently burned or logged areas. Gary Miller and Walter Graul estimated that by the 1980s the prairie sharp-tail's historical Minnesota range had been reduced by 60 percent, and their Wisconsin and Michigan ranges by 90 percent.

Estimated annual hunter kills in Minnesota have been in a long-term decline since as far back as 1949, when 150,000 were taken. Annual hunting seasons have continued since then, with especially low numbers in the mid-1980s and mid-1990s. About 14,000 birds were taken during the late 1990s. State estimates indicate that the sharp-tail population in 2000 was about 70 percent below that of 1980, with the east-central component declining 76 percent and the northwestern population 68 percent. Breeding bird surveys for 1966–1992 also suggest that in Minnesota the population may be increasing in the east but declining in the west. This trend continued from 1996 to 2000, when the eastern range population increased some 5 percent, whereas the northwestern range decreased 9 percent. Overall, the state's population was probably stable during those five years. In the east, 1,106 males were seen on 135 leks during 2000, and in the west 890 males were counted on 118 leks. Such numbers would suggest a spring population of at least 4,000 birds. These favorable counts appear to be the result of an intensive brush control management program by the state's Department of Natural Resources (see www.dnr.state.mn.us). However, in

2001 the statewide population dropped 37 percent, to near-record lows, the lek counts in the east-central region being down 43 percent and those in the northwest 30 percent. In both areas the populations were 76–80 percent below 1980 levels, according to William Berg.

In the early 1930s the Wisconsin sharp-tail population, estimated by Aldo Leopold to be about 55,000 birds distributed in at least 25 central and northern counties, was already rapidly declining. They still occurred in 22 or 23 counties as of 1975, but by 2000 their Wisconsin range had been reduced to fragmented populations scattered across 12 counties. The state's annual kill during the 1970s ranged from about 8,000 to 12,000 birds; fewer than 100 birds were averaged from 1992 to 1995; and since 1997 hunting has been allowed only by special permit. Breeding bird surveys for the period 1966 to 1992 suggest that the Wisconsin population is increasing in the central core of its range but declining elsewhere. Lek counts in 1998 revealed a total of 748 displaying males, suggesting a total state population in the low thousands.

Similar tends can be seen in Michigan's sharptail population and hunter-kill figures. By the early 1980s the annual Michigan kill was below 500 birds, and although kills were permitted in the Upper Peninsula until 1996, they were quite small. Sharp-tail hunting ceased throughout Michigan in 1996, which resulted in a termination of annual lek surveys by the Michigan Department of Natural Resources. However, a survey of the Upper Peninsula by volunteers in 1999 produced 602 birds, of which 262 were males. Most of these were on agricultural lands of the eastern Upper Peninsula and in pine barrens of the Hiawatha National Forest. The area around Seney, in the central Upper Peninsula, was another stronghold. The sharp-tail has now effectively been extirpated from the Lower Peninsula, with 50–75 birds present in 2000. The Upper Peninsula counts in the spring of 2000 tallied more than 400 birds. As in Wisconsin, the Michigan population of sharp-tails was probably not more than 1,000 to 2,000 at the turn of the century.

In these three Great Lakes states a combination of fire suppression, forest succession, the replacement of natural forests with monocultural pine plantations, and modern agricultural practices have all resulted in population setbacks for the prairie sharp-tail. The late 1990s may have seen a slight improvement in the sharp-tail's status in Michigan, owing to increased hay–grassland farming activities in the Upper Peninsula. Breeding bird surveys for the period 1966–1992 also suggest that the eastern Upper Peninsula population is increasing.

PLAINS SHARP-TAILED GROUSE

The plains sharp-tail intergrades in the northeast with the prairie sharp-tail and is separated by the continental divide from the Columbian sharp-tail (Map 8).

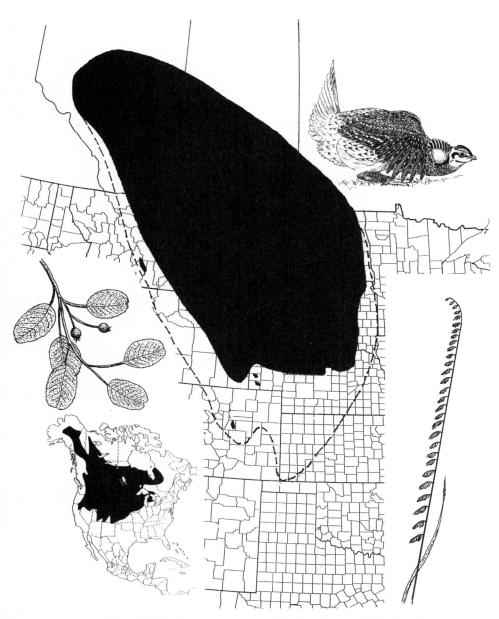

Map 8. Historical (dashed line) and current (inked) distributions of the plains sharp-tailed grouse. The inset map *(lower left)* shows the species' overall historical distribution, including that of subspecies not discussed in this book. Inset sketches illustrate serviceberry *(left)* and side-oats grama *(right),* important food and cover plants of the central plains.

It resembles the latter in being lighter-colored above with less conspicuous white spotting, However, it is larger, grayer, and paler than the Columbian race, with a whiter throat and paler underpart markings. Its historical range extended from the midgrass prairies of east-central British Columbia and southern Alberta southeastward to the Red River Valley of North Dakota, across the glaciated plains of South Dakota, the Nebraska Sandhills, and south to western Kansas and the Oklahoma panhandle. It also extended west across the high shortgrass plains of Montana, Wyoming, and eastern Colorado. It is a race that seemingly prefers shortgrass and midgrass prairies, but with some woody cover present, the latter in a ratio of about 5–30 percent of the total available habitat. Among the shrubby plant species that are used especially for winter foods are the fruits and buds of roses, chokecherries, serviceberries, willows, and poplars.

By 1900 the plains sharp-tail had begun a northward retreat from its southernmost range limits, a trend that continues to the present, coinciding with a gradual reduction in natural grasslands. The birds had disappeared from the Oklahoma panhandle by 1931. There were some pre-1900s reports of the birds in the extreme northwestern panhandle of Texas. They were evidently gone by then or even earlier from western Kansas, the last specific record being for the late 1800s. Attempts at reintroductions, plus a probable range extension by birds in southwestern Nebraska, have resulted in a few records from west-central and extreme northwestern Kansas. As of 2000, only three leks were known to be present in the state, and the population seemed to be dwindling.

Of the remaining four Canadian provinces and five states still supporting plains sharp-tails, Colorado has the fewest birds. The plains race once occurred east of Colorado's Front Range south to El Paso and Kit Carson counties and north to the Wyoming border, but a dramatic decline began in the late 1800s, as ranching modified natural grassland patterns. The plains population continued to shrink, and by the start of the 21st century only Douglas County continued to hold a natural population. This population, located not far from Denver, is under increasing threat from Denver's urban sprawl. A few reports of sharp-tails from northeastern Colorado, in Weld and Logan counties apparently represent birds moving a short distance south across the Nebraska–Wyoming border. In this area greater prairie-chickens also occur, and hybridization between the two puts the tiny sharp-tail gene pool at risk. In Douglas County they now occupy rather atypical habitat for plains sharp-tails, namely a combination of grassy openings and brushy cover comprised of oaks, willows, and the fruit-bearing serviceberry, which provides a valuable source of winter foods. Out on the northeastern plains they occur in a more typical plains sharp-tail habitat, now consisting of a mixture of native and introduced grasses, with scattered deciduous shrubs. As of the late 1980s the estimated Colorado population of the plains sharp-tail was 83 to 228 birds, the males using six leks, all located in eastern Douglas County just east of Pike National Forest. The race was declared endangered in the state in 1976.

Unlike Colorado, Nebraska has managed to retain much of its original sharp-tail range, thanks almost entirely to the 20,000-square-mile Sandhills region. This is a region of mixed grasses and shrubs beautifully adapted to dune life and taller grasses growing in meadows and swales, which together provide a perfect combination of food and cover for these birds. It was to these Sandhills that the birds retreated as farming and ranching eliminated their preferred habitats in the eastern and southern parts of the state. The sharp-tail's current range in Nebraska closely corresponds to the sandiest parts of the state, although in the latter part of the 20th century a northern panhandle population expanded as Conservation Reserve Program (CRP) lands were converted back to grasslands and the Oglala National Grasslands of the same region gradually recovered. The estimated annual hunter kills of prairie grouse during the 1998–1999 seasons averaged about 47,000 birds as compared with about 41,000 during the late 1980s and early 1990s. On a statewide basis sharp-tails comprise about 55 percent of the total grouse harvest, suggesting a kill of around 25,000 birds annually. Leonard Sisson estimated in the 1970s that grouse hunters took about 12 percent of the fall population, which if still applicable would extrapolate to a state population of about 250,000 birds. But like Nebraska's prairie-chicken estimates, no supporting data accompany these figures. Breeding bird surveys for the period 1966–1992 suggest that the Nebraska population is increasing in the northwestern and southwestern parts of its range but is declining elsewhere.

Wyoming has not fared as well as Nebraska with regard to its sharp-tails, having lost more than 50 percent of its historical grouse range. This consisted of sagebrush-dominated steppe with intermixed grasses, a habitat easily converted to farming wherever irrigation was an option. The birds are still relatively common locally, east of a line from Sheridan County in the north to Albany County in the south, but they once occurred west to the Yellowstone region where they were geographically replaced by the Columbian sharp-tail. No population estimates for Wyoming are available, but the birds have been hunted annually, with kills of fewer than 5,000 birds in the 1970s. Estimates of harvests for the period 1995–1998 averaged about 2,500, suggesting a population decline since the 1970s. However, breeding bird surveys for the period 1966–1992 suggest that the Wyoming population has been increasing over most of its remaining range. The only known hybridization between the sharp-tailed grouse and the sage-grouse has been reported from eastern Wyoming and eastern Montana.

The historical South Dakota range of the sharp-tail has declined only moderately, mainly along its eastern border where farming has reduced or eliminated its prairie habitats. Throughout South Dakota the birds still occupy about 40,000 square miles, and use midgrass and shortgrass prairies, especially where brushy thickets are also present. These figures suggest a fall population of at least 400,000 birds, assuming a fall population density of no fewer than 10 grouse per square mile of occupied habitat. The birds are hunted regularly in South Dakota, with an average of 79,000 prairie grouse shot annually between 1994 and 1999. According to Tony Leif, perhaps 75–80 percent of these were sharp-tails, suggesting a sharp-tail kill of about 60,000 birds. In 1979 some 23,000 hunters killed 179,000 prairie grouse (averaging about 8 per hunter), whereas in 1999, 16,000 hunters killed 86,000 birds (about 5 per hunter). Studies done in the early 1970s suggested a South Dakota hunter kill of about 20 to 25 percent of the fall population that would imply a fall population of at least 400,000 birds. No efforts are currently being made by South Dakota biologists to estimate statewide grouse populations.

Limited transplanting efforts have been done by South Dakota biologists in attempts to restore the birds into some lost areas of their range. As elsewhere, these sharp-tail transplants have generally proved unsuccessful. Breeding bird surveys for 1966–1992 suggest that the South Dakota population is increasing in the western and northwestern parts of its range but declining in the south. In 1995 and 1996, 150 sharp-tails from South Dakota were translocated to western Iowa in hopes of reestablishing the species in that state. By 2000 only a few birds were still being seen there.

In North Dakota there have been only negligible reductions in the sharp-tail's overall historical range, primarily along the intensively cultivated Red River Valley. However, its population size has certainly declined precipitously since the late 1800s, when one writer said it was difficult to take a buggy through the fields near Fargo without the risk of running over sharp-tail nests. Like the prairie-chicken, a high point in the sharp-tailed North Dakota population may have occurred then or as late as the early 1900s, but no estimates of their number exist. From 1875 to 1886 there was no daily limit on the number of sharp-tails shot; a limit of 25 prevailed until 1908. The first state laws prohibiting traps, snares, nets, artificial lights, or bird lime went into effect in 1897, and the purchase of a hunting license was also initiated then. Thereafter the total length of the season, the legal hours of hunting, and daily as well as possession limits were all gradually reduced. By 1931 hunting was banned in some parts of the state. That did little to stop the population decline, for the dust bowl days had arrived. Efforts to track annual hunter kills and overall grouse populations using lek counts were begun in the early 1950s. By the 1970s the estimated sharp-tail kill in North Dakota averaged about 100,000 to 150,000 birds, suggesting a state fall grouse population of at least 500,000 to 600,000 birds. Breeding bird surveys from 1966 to 1992 suggest that the North Dakota population is increasing over most of the state but is declining in the southwest and along the western edges of the Red River Valley. As of the year 2000, lek counts indicated that populations in the western parts of the state were thriving, and overall the population had increased about 18 percent from the previous year.

Montana's sharp-tail range has been reduced at least 50 percent since its historical high, the losses largely occurring along the species' western edges, mostly as a result of the state's midgrass prairies having been converted to cultivation. Its densest remaining populations are in north-central Montana near the Alberta–Saskatchewan border and in east-central regions near the North Dakota border. During the 1960s and 1970s the average annual hunter kill was about 90,000 and 95,000 birds, respectively, suggesting a fairly stable population during those decades. These hunter kill figures would indicate a fall population of perhaps 400,000 to 500,000 for that period. Breeding bird surveys for the period 1966–1992 suggest that the Montana population is increasing in the middle third of its state range and in the northeastern part of the state but is declining elsewhere.

Judging subjectively from the information on breeding bird surveys provided by John Sauer and others, it would seem that the plains sharp-tail was relatively stable from 1966 to 1992. Periodic updates of these data that have appeared online since then suggest that most of the sharp-tail's increases have occurred in North Dakota, whereas declines have occurred in Alberta and Saskatchewan. Canadian declines, which were greatest through the mid-1970s, have been more gradual since then, but in the Missouri River plateau region south of the Canadian border there have been distinct population increases since the late 1980s. Probably at least part of this latter improving trend is a result of the federal CRP program that has taken marginal farmlands out of production and replaced them with grass cover.

Based on a survey of state and provincial biologists, Gary Miller and Walter Graul estimated in 1980 that the total plains sharp-tail population might consist of 600,000 to 3 million birds, exclusive of what were then probably relatively small populations in British Columbia and Manitoba. Respondents from Canada and all but two states (Colorado and Wyoming) believed that their sharp-tail populations were going to be able to maintain stable ranges, although most also thought that their regional populations would be decreasing over the next decade. Given the rather fragmentary information from individual states and provinces, it is impossible to judge whether this estimate is still a valid one. Based on the somewhat limited information for Montana, the Dakotas, and Nebraska, it is likely that the U.S. fall population of plains sharp-tails must have numbered more than 1 million birds at the start of the 21st century, with by far the largest concentrations in the Dakotas and Montana. The Canadian population occupies a comparably large area, and its population is likely to be at least as large. In such a case, a collective estimate of 2 million birds in the overall fall population of plains sharp-tails seems fairly conservative.

COLUMBIAN SHARP-TAILED GROUSE

Of all the races of sharp-tailed grouse, the Columbian sharp-tail has suffered the most from human activities. It once extended from east-central British Columbia south through Washington, Idaho, and western Montana to northern California, northern Idaho, Utah, and western Colorado (Map 9). Its primary habitat was the sagebrush steppes and bunchgrass prairies of wheatgrasses and associated grasses and forbs of the Pacific Northwest, where summers are dry and most of the annual precipitation occurs during winter. Most of its historical range lies within the semiarid rain shadow that occurs to the east of the Cascades and Sierra Nevada ranges. Its range is limited on the east by the western wooded slopes of the Rocky Mountains.

At least 90 percent of its historical range in Idaho, Montana, Utah, and Wyoming had already been lost by 1980, and more than half was gone by then from Washington and Colorado. Only in British Columbia, where the majority—perhaps as much as 80 percent—of this race's original range then persisted, could it still be considered fairly secure. As of 1980 its population was estimated by Gary Miller and Walter Graul as 60,000 to 170,000 birds, with perhaps 60–80 percent occurring in British Columbia. In 1989 it was listed as a Category 2 candidate (meaning that insufficient evidence was available for a proposed status) for inclusion in the federal government's list of endangered and threatened wildlife. In 1996 it was formally proposed for recognition as nationally threatened by the Biodiversity Legal Foundation. In 1999 a survey by the U.S. Fish and Wildlife Service suggested that the population then was 34,000 to 70,000 birds, with the majority occurring in Idaho. These figures represent a reduction by more than half of the population estimates made about 20 years earlier by Miller and Graul. A similar review by Richard Hoffman and San Juan Silver in 2000 put the estimated spring sharp-tail population at 60,000 birds, distributed in an area representing only 8 percent of their historical range. They judged that only Idaho, Colorado, Utah, and British Columbia then had spring populations of more than 5,000 birds.

The Columbian sharp-tail is now completely extirpated from California, Nevada, and Oregon. It was gone from California by about 1915, from Nevada by the early 1960s, and from Oregon by the 1980s. Reintroduction efforts were made during the late 1990s in Wallowa County, Oregon, with uncertain results. As of 1999 about 50 of these birds were still surviving there. Reintroduction

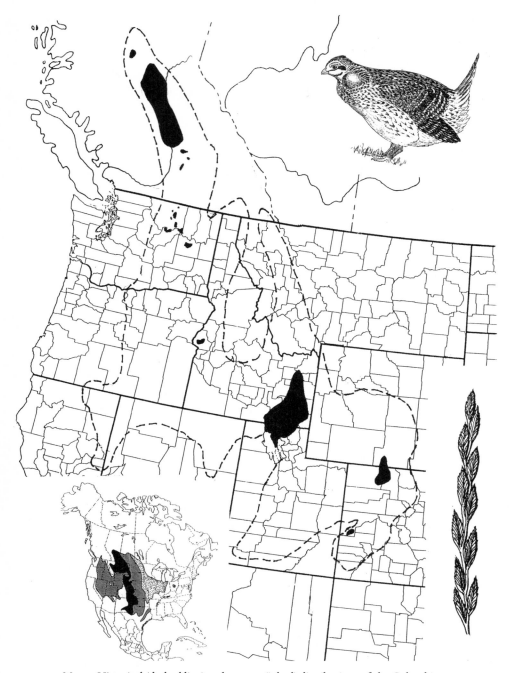

Map 9. Historical (dashed line) and current (inked) distributions of the Columbian sharp-tailed grouse (current distribution partly after Schroeder 2000b). The inset map *(lower left)* shows the approximate presettlement distribution of tallgrass prairie and Texas coastal prairie (stippled), midgrass prairie (horizontal hatching), shortgrass prairie (inked), and sage grasslands (diagonal hatching). The inset sketch *(lower right)* illustrates western wheatgrass, an important sage grassland cover plant.

efforts in Nevada were begun at the turn of the century and have also been made or are planned in California, Washington, Oregon, and Montana.

Colorado, which has a remnant population of plains sharp-tails east of the Rockies, also has remnant flocks of Columbian sharp-tails in the western parts of the state. Gary Miller and Walter Graul mapped four such populations in 1980; as of the late 1990s only one was certainly surviving. This population occurs along the Wyoming border, mostly in Moffat and Routt counties, and occupies sagebrush or similar montane shrubland habitats. Counting a small spillover of this population into Carbon County, Wyoming, this group numbered some 6,000 to 8,000 birds in the late 1990s, making it the second-largest of the surviving U.S. populations. A small flock may also still survive in the vicinity of the Umcompahgre National Forest of Mesa County, since possible breeding was reported there during 1987–1995 surveys for the *Colorado Breeding Bird Atlas*.

In Wyoming the Columbian sharp-tails are nearly gone from the state. Historically they occurred in at least Teton County and the Pinedale region, and they may still survive in small numbers in Yellowstone Park. Their largest remaining population is certainly the one that is located along the Wyoming–Colorado border in Carbon County, where they have been protected from hunting since 1994. The Montana population of Columbian sharp-tails has always been a small one, as they have been restricted to areas of grasslands from about the National Bison Range north locally to the Tobacco Plains near Eureka. As of the late 1990s this remnant state population consisted of only two known flocks of about 50 birds each. By 1999 only 5 birds were seen on a single Montana lek.

Idaho supports the largest remaining U.S. population. It is now mostly if not entirely confined to the southeastern part of the state, with an "overflow" into northeastern Utah. The Idaho population is still considered secure enough to allow annual sport hunting, which as of the late 1990s accounted for about 6,500 birds, This total is thought to represent 10 to 30 percent of the state's overall fall population, or some 20,000 to 50,000 birds. Sharp-tailed grouse have not appeared on breeding bird surveys or Christmas bird counts elsewhere in the state and must be regarded as otherwise extirpated or nearly so from western and northern Idaho. The adjoining Utah population is not so well studied, but Richard Hoffman and San Juan Silver thought that it might number more than 5,000 birds in 2000.

The remnant Washington population of Columbian sharp-tails was once continuous with that of British Columbia, but their range now has been fragmented into eight small flocks in Okanogan, Douglas, and Lincoln counties, primarily around Turk Valley, Dyer Hill, the Colville Indian Reservation, and the Swanson Lake Wildlife Area. These flocks totaled only about 600 birds in

2000, a reduction of 94 percent since 1960, according to Michael Schroeder. By 2000 more than half of 107 leks known to exist in Washington State in 1960 had become vacant. By 1995 these once-common birds of the Palouse bunchgrass prairies and sagebrush steppe had seen most of their native habitats converted to cropland in central and eastern Washington. The native grasslands there were actually reduced by more than 90 percent (from 25 to 1 percent), and sagebrush by more than half (from 44 to 16 percent) of the birds' historical state range. The Columbian sharp-tail was classified by Washington State as a threatened species in 1998.

Gary Miller and Walter Graul estimated in 1980 that of a total Columbian sharp-tail population thought by them to number between 60,000 and 170,000 birds, perhaps 60 to 80 percent of the birds then occurred in British Columbia. This estimate may have been too optimistic, although annual hunter harvests of the 1960s had averaged more than 20,000 birds and might have justified such a conclusion. However, later studies resulted in lower estimates, and the total British Columbian population as of the late 1990s was correspondingly downsized to only 4,500 to 10,000 birds. After receiving a petition from the Biodiversity Legal Foundation to have the Columbian sharp-tail listed for protection under the Endangered Species Act, the U.S. Fish and Wildlife Service concluded in October 2000 that such protection was not warranted.

BREEDING BEHAVIOR AND ECOLOGY

With regard to the greater prairie-chicken, the briefest conservation recipe might be expressed as "grasses, forbs, and grains." The grains component should ideally represent about a third of the total available habitat, with diverse native grasses with an admixture of broad-leaved forbs comprising up to two-thirds. The corresponding shorthand recipe for sharp-tailed grouse might be "grasses, forbs, and shrubs." Thus a maximum of about a third of the habitat should be in the form of taller shrubs or smaller deciduous trees such as aspens and poplars. These provide important roosting, escape, or foraging cover, especially in winter when snow accumulations may hide most ground-level vegetation. The vegetational components should also support an abundance of insects for summer foods, especially during brood-rearing, and provide nesting cover that simultaneously allows for overhead visual protection but enough lateral visibility to detect approaching predators from a safe distance. In more northern climates, densely leaved shrubs with surrounding grasses a foot or so in height often provide such nesting cover. Winter protection from the elements in the form of trees for roosting or providing above-snow food sources may be important, but in deeply drifted areas without such sites sharp-tails will use snow burrows for nocturnal roosting as needed.

There are some racial variations on this general habitat theme. The prairie sharp-tail is more likely to have aspens or willows rather than shrubs as part of its habitat mix. The plains sharptail's shrub component often consists of snowberry, serviceberry, or similar low-stature deciduous shrubs that may provide both physical cover and persistent winter foods. The Columbian sharp-tail is more likely to have more arid-adapted sagebrush species or montane-adapted shrubs such as rabbitbrush and buffaloberry filling this role. Especially in the prairie and plains races, cultivated crops may be locally important sources of fall and winter foods, as with prairie-chickens. In contrast to the situation with prairie-chickens, acorns are only locally significant winter food items (as in Wisconsin); the winter buds and spring catkins of aspens, willows, birches, cottonwoods, and sometimes maples are correspondingly more important.

Another important habitat component is the presence of low-vegetation areas suitable for establishing leks. Lek sites ideally should have sparse or low vegetation, be elevated above the surrounding landscape and offer a panoramic view, and have heavier herbaceous escape and nesting cover fairly close by, within a mile or less. In the more arid parts of the sharp-tail's range these criteria are not hard to fulfill, but the prairie sharp-tail lives amid taller vegetation, and lek locations that may have been formed by burning or cutting tend to regrow back into heavier cover within a few years. Robert Berger and Richard Baydack found that over a 10-year period 7 of 12 prairie sharp-tail leks in the interlake region of southern Manitoba were abandoned as aspen forest cover increased beyond 56 percent coverage of the total area occurring within a 1-kilometer (0.6-mi) radius of the lek site and as prairie cover was reduced correspondingly to less than 15 percent. In another aspect of this study, Baydack observed that active leks typically consisted of 68 percent grass cover, 15 percent each of forbs and bare ground, and only about 1 percent of shrubs. The leks were on sites higher than most surrounding terrain and were usually situated on essentially level ground. Escape cover and trees were located within about 600 yards of leks, which were spaced out at an average interlek distance of about 1.4 miles. As with the leks of prairie-chickens, this spacing probably allows female sharp-tails to be within sight or sound of at least one, and possibly more than one, of the leks located within their home ranges.

Because the requirements for prairie sharp-tail leks are so similar to those of greater prairie-chickens, it is not unusual for the two species to share a single lek in those areas where they coexist. Such mixed leks are unlikely to have an equal number of participating males; instead only one or two males of the less common species are typically present, which in Nebraska is most often prairie-chickens. Mixed leks are also less common where males have a chance of joining a conspecific lek that is nearby. This is the case in the eastern Sandhills of Nebraska, in eastern North Dakota, and in Wisconsin. It was also true on Manitoulin Island, in Lake Huron, until the prairie-chicken population there eventually disappeared, probably at least partly because of competition with the more common sharp-tails.

In such areas hybridization between the two species sometimes occurs, usually at a rate that does not exceed 3 percent of the combined population. However, the hybridization rate was found to be much more common in a few areas of Nebraska where the two species had only recently come into contact. The first-generation hybrids are essentially intermediate in appearance and also in their male displays (Figure 16). They are also known to be fertile, and sometimes even produce backcross offspring. However, in my experience hybrid males never manage to obtain the desirable interior territories and are usually attacked by both parental species. Hybrid females are unlikely to be recognized as such, and thus probably are not rejected for mating by males of either parental species. It may well be that it is the hybrid females who are responsible for producing most or all second-generation hybrid offspring.

Like prairie-chickens, sharp-tails no longer migrate as far as they once did to avoid extreme winters, thus allowing males to remain much closer to their leks throughout the year. The adult males typically revisit their lek sites in September and October, and may even engage in some display during this fall period. Juvenile males may thereby at least learn lek locations, but they are unlikely to participate in actual displays at that time. Leks are probably abandoned by the first snowfall, but the males remain nearby and may begin to visit the leks again during warm days in early March or even February, as day lengths begin to in-

Figure 16. Hybrid sharp-tailed grouse × interior greater prairie-chicken, male in booming/cooing display posture (after photo by Ed Bry).

crease significantly. As the number of males appearing on the lek increases, the fights and the level of displays increase correspondingly, as experienced males try to regain their preferred sites or attempt to move into favored locations made available by the death of males that had held interior territories.

Richard Baydack found that 13 percent of the males on leks performed 93 percent of all observed copulations; most of these matings were achieved by males holding central territories. They had attained these positions through overt aggression or by moving in from peripheral territories they had held as yearlings the prior season. Baydack believed the lek more closely resembles a true territorial system than a strict dominance hierarchy. A similar restriction of mating to only a few dominant birds was reported by R. Evans, who noted that 20 of 21 matings (95 percent) were attained by a single male. Harry Lumsden noted that in one lek a single male accounted for 76 percent of all observed matings.

Characteristics that distinguish dominant males from other males are of special interest. Michael Gratson found that most of the variation in observed mating success by males could be explained by the successful males' having longer dancing times, faster tail-clicking rates, and shorter intervals between "cork" calls. Indirectly, smaller body size (which may help facilitate more energetic or prolonged displays, as has been noted in sage-grouse) and a central territorial location were also significant elements not specifically selected for by females. Females evidently chose mates that provided high-intensity stimuli inducing early mating, possibly providing both direct genetic benefits and also "economic" benefits (such as temporal mating advantages) to these females. Gratson found that nine of 47 permanently established males on four different leks performed 75 percent of 72 observed copulations.

In contrast to these findings that seemingly favor smaller body size but more active behavior, Leonard Tsuji and others reported that central territorial males, all of which were adults, tended to be larger in body mass than peripheral ones. The latter group was comprised of mixed adults and immatures. The larger body mass of the more central males probably helps them dominate competitive interactions associated with territorial acquisition and maintenance involving other birds. The central males also had larger average testis volumes, which is probably related to relatively greater available sperm quantities. Other studies have shown that males holding central territories not only have higher levels of mating success but also possess more motile sperm than peripheral ones.

The male displays of sharp-tailed grouse are sufficiently different from those of greater prairie-chicken to have engendered some distinctive names. The best descriptions are those of Harry Lumsden, who first named most of the displays, and Ingemar Hjorth, who provided more detailed analyses of postures and sounds, and renamed many of the displays. The most conspicuous and animated of all the sharp-tail's displays is dancing (sometimes called "tail-rattling," as by Harry Lumsden). This complex display (Figure 17) begins with an animated phase, during which the wings are extended almost horizontally, the tail is cocked vertically, the head is somewhat lowered, and the esophageal air sacs of the neck are only slightly expanded, exposing their purplish skin. The yellow eye-combs are expanded but are always smaller and less evident than those of prairie-chickens. After assuming this posture the bird begins a rapid foot-stamping, resulting in the same drumming sound that is produced by prairie-chickens before booming, but it is more prolonged, often lasting 4–6 seconds. Instead of remaining stationary as do prairie-chickens, the bird advances forward like an animated toy airplane, and instead of moving in a straight line is likely to move in arcs or tight circles. At the same time, the tail is shaken from side to side as a result of and in synchrony with the foot-stamping, both at a constant rate of nearly 20 per second. The sound produced by the tail feathers rubbing past one another is a dry rattling or clicking sound similar to that of mechanical clockwork; the foot-stamping is much lower-

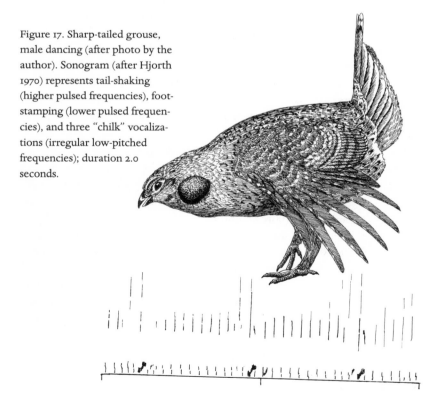

Figure 17. Sharp-tailed grouse, male dancing (after photo by the author). Sonogram (after Hjorth 1970) represents tail-shaking (higher pulsed frequencies), foot-stamping (lower pulsed frequencies), and three "chilk" vocalizations (irregular low-pitched frequencies); duration 2.0 seconds.

pitched. However, from even a short distance the sounds merge into a soft whir, and their collective noise is usually inaudible beyond a few hundred feet.

During the dancing phase the male sometimes utters a repeated, rather shrill squeaking sound, called a "chilk," or a somewhat softer "cha" note. This is a relatively far-carrying call, audible up to several hundred yards away. A different note, called the "cork" call because it sounds like the noise made when a cork is pulled out of a bottle, is a soft, brief (.06-sec) call and is uttered less frequently, but may be produced up to six times in a single dancing sequence. It is especially prevalent when females are on the lek. It may have high-pitched components (up to 2,400 kHz according to Henry Kermott and Lewis Oring) or be quite low pitched (under 1,000 kHz in Hjorth's sonograms), and it may have as many as three sound components.

The second phase of dancing is the "relaxing" phase, sometimes called "posing" or "freezing." The bird suddenly stops in place, lowers and partly retracts its wings, and adopts a rigid posture while uttering a few soft, low-pitched cooing or crooning notes. This phase often lasts several seconds, then the animated phase suddenly resumes. It is not unusual for several males or all males on the lek to almost perfectly synchronize their animated and relaxing phases of the dance display, making the performance resemble a highly choreographed routine. It has been suggested that the most active bird in the arena is most likely to initiate a dancing bout, but the almost perfect synchrony of all the males makes this point difficult to substantiate. During these periods of near-silence the birds may be able to detect either aurally or visually any unusual activity in the vicinity, such as the approach of potentially dangerous intruders. The animated phase of the dance may last up to eight seconds during intense displays, and the intervals last approximately equal periods. When females are present on the lek, the frequency of dancing greatly increases; the dancing birds strongly orient toward them; and the dancing sequences are of longer durations.

If a female is near a dancing male, he may attempt to approach her from behind, continuing his foot-stamping and tail-rattling until the moment he tries to mount her. Or he may pause and adopt a semiprostrate, or "nuptial bowing," posture before her, exactly like that of greater prairie-chickens (see Figure 12, p. 75). He will remain for several seconds in this crouched position with his wings partly extended and his bill touching the ground. This posture, which is not an invariable precursor to copulation, is rather infrequently observed. Copulation takes the same form as in prairie-chickens and other grouse, lasting but a few seconds. There is fairly good evidence to support the view that a female copulates only once before starting her clutch: the reproductive tracts of females entering leks have not registered any traces of sperm.

One other important postural display and call is of special interest because of its similarity to the booming display of prairie-chickens. This is the "bowing" and "cooing" posture and call (Figure 18). From an "oblique" body posture, with tail cocked and head somewhat raised, the male moves his head forward and downward, simultaneously inflating his neck pouches (air sacs) so that they swell outward, but not to the degree found in prairie-chickens. The crest feathers are raised somewhat, and the eyes are nearly closed, as a single- or double-noted cooing sound is produced, much like that of a dove or owl. In many cases a gobbling rather than cooing note is uttered, having from one to four syllables, and the neck skin is only slightly expanded. This "gobbling call" seems to be a low-intensity version of the cooing call. Both gobbling and cooing play much smaller roles in sharp-tail displays than does booming for prairie-chickens; they seem to be low- to moderate-intensity aggressive displays directed toward other males, rather than having combined aggressive and sexual connotations.

Figure 18. Sharp-tailed grouse, male bowing and cooing (after various sources including photos by the author). Sonogram (after Hjorth 1970) shows calls by two males (calling sequence A–B–B–A); duration 2.0 seconds.

Figure 19. Sharp-tailed grouse, male in wide-necked upright posture (after photo by Harry Lumsden).

A few other displays are worth mentioning. Flutter-jumps are rare in sharp-tailed grouse and tend to be replaced by on-the-ground wing-beating. Males sometimes adopt a "wide-necked upright" posture (Figure 19) with their bare neck skin exposed and the feathers lying directly above it raised vertically to form a distinct ridge. This posture is often assumed prior to adopting other postures, such as dancing or the oblique. Males spend a great deal of time at the edges of their territories, crouching and uttering rapidly repeated cackling or grunting notes similar to those of prairie-chickens (Figure 20). Fights sometimes break out at these territorial boundaries and take a form exactly like those of prairie-chickens. At other times two birds may perform "walking or running in parallel," with their heads withdrawn and their wings extended and drooping, the two occasionally stopping to turn and face one another. Hjorth called this display the "parallel forward advance" and considered it as only a variant of a simple hostile forward advance by both birds to a point of territorial boundary confrontation.

BREEDING BIOLOGY

From the time a female is fertilized on the lek to the laying of her first egg is evidently quite short, probably 1–3 days. The peak of mating activity is similar to that of prairie-chickens in the same region, typically from as early as mid-April in southern parts of the range (as is typical in Nebraska) to early May in more northerly regions. However, yearly variations in weather can have delaying or facilitating effects, which in North Dakota may be as much as two

Figure 20. Sharp-tailed grouse, males in territorial confrontation (after photo by the author). Sonogram of cackling after Hjorth (1970); duration 1.5 seconds.

weeks. Additionally, smaller late peaks in mating may occur as a presumed reflection of renesting efforts by unsuccessful females, as also occurs in prairie-chickens.

Given the usual short interval between copulation and egg-laying, it is possible that the females choose their nest sites in advance of mating. Although the nest may not be located closest to the lek where mating occurred, the average distance in various studies has ranged from 0.25 to 1.1 miles, or less than the usual average interlek distance. Further, experienced females often nest near the location of their previous-year's nest, perhaps depending in part on the relative success of that nesting effort. More than one renesting effort may be made by females suffering failed clutches. A maximum of four nesting attempts has been documented.

Anthony Apa tested the role of nest cover in nesting success in Columbian sharp-tails. He radiomarked 48 female sharp-tails and followed their nest success. All the females that nested in native vegetation had successful nests, whereas only 45 percent of those nesting in nonnative vegetation did. Additionally, through the use of 551 artificial nests, he found that nests located farther from leks were more successful than those placed near leks. His result supported using a male-avoidance strategy for successful nest-site selection, rather than a sentinel-decoy model, namely placing nests close to male leks in order to benefit from their presence as potential sentinels.

In a somewhat related study, Douglas Wachob investigated the role of the

Conservation Reserve Program on nesting plains sharp-tails in Wyoming. In that area CRP lands comprised 15 percent of the study area, with highly grazed rangeland and winter wheat occurring in most of the remainder. All the sharp-tailed nests found were in CRP land. Further, hens with broods used CRP patches having a high coverage of weeds and annual grasses as well as irrigated alfalfa patches relatively more often, and non-CRP patches less often, than was expected by their relative availability. He also found that the relative incidence of CRP lands in his study area correlated positively with both the number of sharp-tail leks present in the study area and the number of males attending these leks. He concluded that CRP lands are vital to reproduction of sharp-tails in southeastern Wyoming. Significant increases in sharp-tail populations in the adjacent panhandle of Nebraska have occurred since the CRP program began, suggesting its importance there too.

Like prairie-chickens, female sharp-tails lay their eggs at the approximate rate of 1 egg per day, until their clutch of 10 to 13 eggs is complete. The average for a variety of locations and studies was 11.8 eggs. Second clutches in Manitoba were essentially of the same mean size as first clutches (11.6 eggs among 14 clutches), and third and fourth clutches averaged only slightly less. Second nests in a Colorado study by Jennifer Boisvert and others averaged about 2 eggs less than did first clutches (7.7 vs. 10.1 eggs).

Incubation usually begins with the laying of the last egg, but there is some variation in this. Fertility and hatchability are typically high, with hatching occurring at 21 to 23 days. Most nest losses occur as a result of predation. Nest success rates are quite variable, but often average around 50 percent. In the Colorado study, nest success averaged 48 percent; more than 60 percent of the successful females were still leading broods; and about half their chicks were still alive. Most brood losses occur at early ages, when the young are sensitive to cold, rain, and a lack of insect foods. Brood sizes often average about half that of clutch sizes, so there might be at least a 50 percent early mortality rate. Typically the young birds remain within a mile or two of the nest site while broods are still intact, but fall dispersal may take them several miles from their natal homes.

Sharp-tails appear to have high annual mortality rates, especially in heavily hunted populations, where they may range from 58 to 83 percent. Such high rates, especially those of more than 60 percent, seem excessive and might serve as a warning to game managers that the population is declining. In one non-hunted Washington population the estimated annual mortality rate was 47 percent, which is probably fairly typical for gallinaceous birds of this size.

6

DARK SHADOWS IN SILVER SAGE
Enigmatic Grouse of the Sagebrush Steppes

Trying to write a book on the prairie grouse of North America without mentioning the sage-grouse is something like trying to interpret native prairies without recognizing the role of big bluestem—it is quite difficult. Although one might argue that sage-grouse are better classified ecologically as desert-dwelling rather than grassland-adapted grouse, these mostly sage-eating birds do interact locally with the much smaller prairie grouse on the shortgrass plains and even on occasion hybridize with sharp-tails. However, they are not believed to be close evolutionary relatives of the typical prairie grouse, having perhaps descended instead from some ancestral forest-dwelling and *Dendragapus*-like grouse that have adapted to arid scrub environments. Or the blue grouse might be a forest-adapted species that was derived from and is most closely related to the prairie grouse group, according to Vittorio Lucchini and others.

To add to these taxonomic complexities, we now know that, instead of only one species of sage-grouse as had long been believed, there are actually two. The Gunnison sage-grouse of southwestern Colorado, identified in the early 1990s, is the first truly "new" bird species to be discovered in North America for more than a century. That fact alone makes the two species of sage-grouse of special interest. Furthermore, like the prairie grouse and in contrast to the forest grouse, both the range and numbers of all sage-grouse populations have been seriously impacted by human activities during the past century. As a result, at least the newly discovered species, and perhaps both species, will likely soon follow the typical prairie grouse into the category of threatened or endangered North American birds.

When Lewis and Clark discovered the greater sage-grouse during their expedition to the Pacific Northwest, they called it the "cock of the plains." It was not formally named and described until about two decades later by Charles Bonaparte, eldest son of a younger brother of Napoleon who had immigrated to America after Napoleon's defeat at Waterloo. For several years he worked at the Philadelphia Academy of Arts and Sciences where he named various species of birds and for his efforts had the Bonaparte's gull named in his honor.

The greater sage-grouse is easily the largest grouse in North America. Its size and mass are exceeded only by two considerably larger Eurasian species of forest-adapted capercaillies (*Tetrao* spp.). The ecological significance of this large body mass among grouse is not at all obvious. Like the capercaillies, sage-grouse consume large quantities of foods low in available energy relative to intake volume. Their large size may allow for great quantities of food intake, or it may somehow be related to interspecies variations in nutritional levels. A somewhat comparable situation occurs in Africa, where a group of ground-dwelling and -grazing birds called bustards is an approximate ecological counterpart to the grouse. There, the largest species are not the ones living in the lushest grasslands but are instead the 15- to 25-pound Arabian and kori bustards (*Ardeotis* spp.) ecologically associated with desert and semidesert steppe environments. In the largest grouse, as in these largest bustards, sexual selection favoring powerful, highly conspicuous males that can attract females over great distances and can dominate potentially competing males has evidently been the driving force favoring larger male body size. The body mass of female sage-grouse, roughly half that of males (Figure 21), is not much greater than that of the other prairie grouse.

Unlike the forest-dwelling capercaillies, sage-grouse are primarily brush-dwelling species of semideserts and are specifically adapted to sagebrush, a large genus (*Artemisia*) of mostly long-lived woody plants occurring in semi-arid to arid parts of North America. Of the several dozen species of sage occurring in western North America (Utah alone has 24), big sagebrush (*A. tridentata*) is the most widespread and easily the most important to sage-grouse. It is part of a multispecies portion (Tridentatae) of the large shrub genus *Artemisia* that occupies more than 265,000 square miles of the intermountain region of western North America, and the range of *tridentata* itself closely corresponds with the historical range of the sage-grouse. Sage-grouse prefer to eat certain species or varieties of *Artemisia* over others, and in some areas the local population of big sagebrush may not be the preferred sage food, whereas in other areas it is preferentially eaten. Such preference differences suggest that geographic differences in palatability may exist, probably owing to differing kinds or concentrations of the plant's essential oils, especially terpenes and related phenols. Essential oils are those plant products that produce distinct,

Figure 21. Greater sage-grouse, male *(right)* and female (after photos by the author).

volatile odors or flavors but are not necessarily essential to the plant's physiology. It is likely that these distasteful substances in sagebrush have evolved to reduce or prevent leaf consumption by insects or vertebrate grazers. Sagebrush species as a group are collectively notable for their associated aromatic odors and acrid tastes; vermouth is a well-known sage-based alcoholic derivative, as is absinthe. A widespread European species of *Artemisia* called "wormwood" was so bitter that it was traditionally believed to provide a purgative cure for intestinal worms.

Sage-grouse not only utilize sage leaves, flowers, and buds as their most important single year-round food source but also use the plants for nesting, roosting, and escape cover. Some winter-hardy species such as big sagebrush seasonally produce two different kinds of leaves. One type is the larger and more tender leaves of summer, during which time adult sage-grouse supplement their basic diet of these leaves and flowering parts with those of many other plants, especially legumes. The sage's summer leaves are gradually replaced during fall by smaller, persistent, and more cold-tolerant leaves that often provide the only above-snow food source for these heavy-bodied grouse during winter. Sage leaves in winter are high in digestible protein, averaging even higher than alfalfa, and as such, where sage is common the leaves are one of the most important winter foods for native grazing and browsing mammals such as antelope.

Perhaps because of the high cellulose content of sage leaves, the digestive tracts of sage-grouse are unique. Instead of a typically highly muscular avian gizzard as in other grouse, the stomach is larger, less muscular, and rather saclike, resembling those of raptorial birds. The digestive tract is supplemented by two extremely long (up to about 30 inches) intestinal pouches (ceca) located at the end of the small intestine. These ceca are collectively longer than the small intestine itself, probably allowing for some gradual digestion of cellulose or lignins through bacterial fermentation processes. Other grouse species such as the typical prairie grouse also have prominent intestinal ceca, but they are much shorter. Sage-grouse males are also notable for having paired esophageal pouches (air sacs) that can be inflated (with air from the true air sacs of the respiratory system) up to about 50 times their normal nonbreeding-season volume. However, this adaptation is related to territorial advertisement display, and is not concerned with digestion, as the food-storing crop is located some distance behind the inflatable portion of the esophagus.

Sage-grouse exhibit the highest degree of sexual dimorphism in adult size and body mass (see Figure 21). Adult males of greater sage-grouse typically weigh 1.7 to 2.0 times as much as adult females, and in the Gunnison sage-grouse the usual adult male to female mass ratio is about 1.75 to 1.0. This compares with a male-to-female mass ratio of 1.2 to 1 in prairie-chickens and 1.3 to 1 in sharp-tailed grouse weighing little more than 2 pounds, but as much as 2.0 or 2.3 to 1 in the European capercaillie *(Tetrao urogallus),* whose adult males may weigh 8 to 10 pounds. It would seem that the larger the grouse, the more disparate the body mass ratio between males and females. One possible explanation for this trend of increasing sexual dimorphism with increasing body mass might be a result of differential growth rates or durations in the two sexes, producing an accelerating increase in body mass with only a moderate increase in body size. It may also be that under the most intense pressures of sexual selection, as occurs in the very large leks typical of sage-grouse, it is not the most actively displaying male but rather the strongest male that achieves dominance over other males and successfully attracts females. Such intense male-to-male competition is clearly obvious in sage-grouse but is not nearly so apparent in the forest-dwelling capercaillies. In these species the competing males are rarely within sight of one another, and females are likely to have only one potential mate in their field of view at any one time.

GREATER SAGE-GROUSE

We will never know with any certainty just how many greater sage-grouse occurred in North America during the presettlement period, but by the 1920s and 1930s the species was already in serious decline across much of its range. In 1985

Clait Braun made an educated guess that 2 million might have existed as of 1950. I estimated from rather indirect evidence that the species' total fall population numbered perhaps 1.5 million birds in 1970, of which about 250,000 were then being shot annually by hunters. At that time hunting was still allowed in 10 states and in 1 province.

During historical times the sage-grouse's geographic range has decreased by at least a third (Map 10), seemingly in close correlation with an approximate 30 percent reduction in the range of sagebrush during the last three decades of the 20th century, especially the range of big sagebrush (*Artemisia tridentata*, including variant types). According to Clait Braun, as much as a 30 percent overall population decline also occurred between 1985 and the late 1990s, to about 142,000 adults. According to Sara Oyler-McCance and others, there was an overall average 20 percent loss of sage habitat in southwestern Colorado between 1958 and 1993, although the rate of habitat loss in the Gunnison Basin was substantially less.

Assuming a 1950 greater sage-grouse population of perhaps as many as 2 million birds, its estimated total population has been reduced by about 90 percent in only five decades. States having more than 20,000 breeding birds in 1950 were Montana, Wyoming, and Oregon. Idaho and Nevada had fewer than 20,000 each; Colorado and Utah had fewer than 15,000 each; California, fewer than 5,000; the Dakotas and Washington, fewer than 2,000 each; and the Alberta and Saskatchewan populations each consisted of about 500 birds. The populations of Washington, Oregon, and (formerly) southern British Columbia were described in 1946 as a distinct, slightly darker and smaller subspecies, sometimes called the "western sage-grouse." However, the race was based on only 11 specimens, and doubts have been cast on its validity. This taxonomic puzzle is now a factor in the conservation battle. It might be easier to get a recognized subspecies listed as endangered than a population simply isolated from all others.

At the turn of the 21st century, hunting was prohibited in Canada but was still regularly permitted in 9 states (Montana, Wyoming, Colorado, Idaho, Utah, North Dakota, Nevada, Oregon, and California). In 2000 South Dakota instituted a two-day season, the first since 1970. Wyoming has traditionally had the largest hunter harvests of any state, averaging more 50,000 annually in the late 1960s and more than 83,000 in the late 1970s. From 1995 to 1998 the kill averaged slightly under 14,000 birds each year.

The historical range of greater sage-grouse was reduced from 16 states and 3 Canadian provinces to 11 states and 2 provinces by the year 2000. States in which sage-grouse populations have been extirpated include Arizona, New Mexico, Oklahoma, and Nebraska, although those that occurred in these states might have represented the newly described Gunnison sage-grouse rather than the northern. It is doubtful that the species ever occurred in Kansas.

Map 10. Historical (dashed line) and current (inked) distributions of the northern and Gunnison sage-grouse. The probable historical range of the Gunnison sage-grouse lies within the area indicated by cross-hatching. The inset map *(lower left)* shows the historical range of big sagebrush, a stem of which is shown at lower right.

Marginal populations of the greater sage-grouse at great risk to extirpation exist in the western parts of the Dakotas and in central Washington. A more secure population occurs in Oregon. These two latter states support the only remaining component of the rather poorly defined western race. Malheur, Harney, and Lake counties in Oregon supported an estimated 24,000–58,000 birds as of the 1990s, with an additional 3,000–8,000 birds in Baker, Crook, Deschutes, Grant, Klamath, Union, and Wheeler counties. About half the Oregon range was lost between 1900 and 1950; since then the rate of range loss has declined. Closed hunting seasons have been periodically established since 1976, and harvests during open seasons have been controlled by limiting permits and tags. In Washington the sage-grouse has been extirpated from 7 counties and as of 2000 occupied only about 10 percent of its historical range, with a total population of about 1,100 birds. More than half these birds occur mainly on private and state-owned lands in Douglas County; the rest are located on the Yakima Training Center, in Kittitas and Yakima counties, which is controlled by the military. Lek counts in 2001 at the Yakima site revealed 276 males at nine leks, down 9 percent from 2000. In July 2001 the U.S. Fish and Wildlife Service determined that this population warranted protection under the Endangered Species Act, but such listing was precluded because of other species' having greater need for protection. In a scale of protection needs ranked from 1 to 12, the population was given a priority rating of 9, with 1 having the highest priority. The total number of such candidate species was then 236. In Canada, the British Columbian population of this western race is now gone. In Alberta and Saskatchewan the northern subspecies' remaining population declined 66 to 92 percent from 1970 to 2000, and calculations by Cameron Aldridge indicated that if current trends continue the Canadian population will likely drop to numbers too low to support a self-sustaining population within two decades. Low rates of chick survival may be the most important factor now influencing sage-grouse survival in Canada.

Wherever sage-grouse still occur, they are never found far from sagebrush, especially during winter. They are highly mobile during that period and may range over areas as great as 50 square miles. Even such large areas may not be large enough; one migratory band in Washington State was estimated to use perhaps as much as 680,000 acres of territory. During winter the local distribution of greater sage-grouse is closely linked to the presence of sagebrush plants that reach a foot or more above snow levels and where the total sage cover represents a substantial percentage of the overall plant canopy. In addition, judging from work by T. E. Remington and Clait Braun, the birds are somehow able to identify and select sage stands having the highest protein levels. They may escape the coldest temperatures by tunneling into the softest snow on the lee sides of sages or other tall shrubs or, if the snow is deep enough, tunnel into snow drifts on more open areas lacking shrub cover above the snow surface.

As winter wanes, the birds begin to move into different habitats. The females search for sage uplands having an understory comprised of a diversity of broad-leaved forbs that are rich in nutrients such as calcium and phosphorus and high in protein levels. By nesting time such areas are also likely to have an abundance of insects, a critical component of potential breeding success. At the same time, males begin to congregate around potential lek sites, gathering along exposed ridges, knolls, and other habitats that tend to be kept snow-free by winds. Such habitats may include not only natural sites but also variously disturbed areas such as recently burned sites, roads, gravel pits, or grassy landing strips. Here sage is low or perhaps even lacking. The presence of nearby sage cover is also important, and males may preferentially gather in areas where females are prone to congregate before nesting—evidence supporting the so-called hotspot model of lek formation. An opposing model, the hot-shot model, supposes that females are attracted to the location where the most attractive males are to be found, regardless of their own habitat preferences.

In contrast to the leks of typical prairie grouse, sage-grouse leks tend to have more males present. There might be several reasons for this. The large size and great mobility of the birds may make it worthwhile for males to travel considerable distances to participate in the single best site relative to female attraction. Since sage-grouse have the highest degree of sexual dimorphism of any of this group, it seems likely there is also a clearer mass-to-fitness ratio among males than in species having only slight variations in male mass and associated dominance potentials. That is, the greater degree of size and dominance variation among the variously sized males in a local population may make it possible for a single alpha male to attract more females, thus making it advantageous for a greater number of males to associate with that particular male, resulting in larger lek sizes.

In any event, it is not uncommon for a sage-grouse lek to attract several dozen males, and in earlier days, such as the mid-1900s, up to a few hundred males were sometimes documented on single leks. At some point, these "megaleks" probably become impracticably large, as no single male could possibly dominate that many males and still find time to fertilize all the females attracted to it. It was the lek behavior of sage-grouse that first resulted in coining the term "master cock" to describe the alpha-rank male who typically obtained the majority of the total copulations on grouse leks, or at least more matings than any other single male. In several different studies, the alpha male has been found to be responsible for anywhere from 47 percent (R. Wiley) to 61 percent (J. Hartzler and D. Jenni) of the total observed matings. In the latter study, a single male was observed to obtain 169 copulations during a single season, and it is not unusual for the dominant male to copulate as many as 20 times in a single morning. Immediately peripheral males, sometimes called "subcocks," are responsible for most of the remaining matings, and the more peripheral males, called "guard cocks," typically obtain few if any. These outermost birds are typically young, inexperienced, and perhaps smaller and less physically fit males. One master cock in the study by Hartzler and Jenni was at least three years old; the other males that were observed to obtain copulations were also from the older age groups.

As with the other prairie grouse, the means by which females can identify the fittest male on a lek are not always obvious. Beside choosing males that are centrally located, females also appear to favor males that are clearly dominant, are most active in their display behavior, and possibly those that are visually or acoustically most attractive. Females may be able to assess the male's relative overall vigor, including its health or parasite load, indirectly by using some of these same criteria, and may engage in "copying" behavior, namely choosing to mate with a male that has already been selected by one or more other females. It is not unusual on sage-grouse leks to see a group of females clustered closely about a single dominant male, waiting their turn for mating, as is typical of Eurasian capercaillies *(Tetrao urogallus)*.

The first males to arrive on the leks in late winter are the older, experienced males. They soon take positions similar to if not identical to the ones they occupied during the prior display season, and they fight fiercely over any vacant spaces made available by the deaths of prior territory-holders. Fights usually involve wing-beating of the opponent, as well as some pecking or prolonged biting, but the males perform jump-fighting with attempted clawing far less often than do the smaller and more agile prairie grouse. Territories within sage-grouse leks are relatively small as compared with those of typical prairie grouse. They may be as small as about 6 square yards, or more than 100 square yards. As a result, males on adjoining interior territories may be only 10 feet apart, whereas some on the periphery may be more than 100 yards apart.

Male sage-grouse on their leks often begin their so-called strutting displays well before sunrise; at that time their exposed and reflective white breast feathers, as well as the rapid inflation and sudden deflation of their anteriorly oriented esophageal pouches produce a curious visual "blinker" effect in the half-light, which adds to the surreal nature of the display. Their tails are held vertically cocked and spread, with the long, white-tipped under tail coverts also widely spread, resembling a sea-urchin's spines when seen from the rear. The white breast area is kept partially inflated, producing a sort of pendulous pouch, the sides of which are lined with unusual feathers that have reduced vanes and strongly stiffened shafts (Figure 22). These shafts are directed down-

Sage-Grouse

Figure 22. Greater sage-grouse, male in strutting posture (after photo by the author). Also shown *(bottom left)* are a filamentous ornamental filoplume from the upper neck *(far left)*, a stiffened white breast feather from the lower neck *(lower right)*, and a greater primary covert *(upper right)*, and *(above)* a lateral tail feather (all from specimens, to scale).

ward and somewhat posteriorly, so that when the folded wings are brought forward the leading edge of the wing rubs over them "against the grain," and a distinctive scraping noise, called "wing-swishing," is produced.

A bout of strutting (Figures 23 and 24) lasting about 2 seconds begins with the male in an upright posture, with the wings held downward at the sides but outside the flank feathers. The white neck feathers are raised, exposing the airy filoplumes lining the sides of the neck and producing an antenna-like array of these feathers behind the head in a sort of semihalo. The yellowish olive eye-combs are only slightly expanded and unlike those of prairie grouse are not very conspicuous. The start of strutting is marked by a rapid wing-swish and a vertical upward jerk of the head, this phase lasting about 0.2 seconds. There is then a pause, as the esophageal pouch is inflated and two bare, breastlike areas of frontal, olive-colored skin are partly and briefly exposed. Simultaneously with the wing-swish noise is a low-pitched "growl," apparently produced by feather noise rather than vocally. After a slight pause there is a second wing-swish, accompanied by a second vertical head-jerk and a greater esophageal expansion, that produces a second brief but silent exposure of the bare skin patches. At this point the male utters a quick series of three low-pitched cooing or hooting notes. There is then a hollow plopping sound as the now fully expanded and orange-shaped air-sacs are deflated. This in turn is immediately

Figure 23. Greater sage-grouse, male strutting display (after photo by the author).

followed by a sharp whistle and a second extremely rapid sac inflation and deflation plop. At that point the bare skin areas gradually disappear, often in a pulsating manner, and a few soft hooting notes may be uttered.

This entire complex sequence of wing movements, head movements, vocalizations, and esophageal expansion and deflation occurs so quickly that it seems to be almost a mechanical action, resembling in Ingemar Hjorth's words "a Morse flashing with an inherited code." There is little similarity between the strutting of a sage-grouse and the advertisement displays of prairie grouse such as sharp-tails. Wild hybrids between sage-grouse and sharp-tails are of intermediate size and appearance (Figure 25), but their male displays consist of a rather rudimentary version of rapid wing-lowering and partial esophageal inflation, with no clear similarities to the displays of either parental species. Except during strutting displays, sage-grouse are silent; they lack the chattering, whining, and yelping noises so common during the territorial advertisements and aggressive encounters of the prairie grouse. Male sage-grouse are more like the capercaillies, in which protracted posturing and contrasting plumage patterning are more conspicuous than are their rather soft vocalizations and mechanical sounds.

Lek displays in the sage-grouse begin during late winter and reach a peak when most copulations are occurring. This typically happens during March or April, depending mostly on the region but with weather having a small timing influence. Young females may collectively copulate somewhat later than adults, and renesting birds may stretch the copulation period out for a month or even longer. A maximum of two attempted renests have been documented, with adult females more likely to renest than yearlings.

Figure 24. Greater sage-grouse, male in strutting display sequence (sketches A–H); sketch I shows rear view of male during entire sequence. Numbers indicate time in seconds since start of display (adapted from Hjorth 1970). Lateral threat between two territorial males is also shown (J), with more dominant male on left (after photo by the author).

Figure 25. Hybrid sharp-tailed grouse × greater sage-grouse, male in strutting posture (after photo by William Strunk).

It is believed that some females may select their nesting areas well before copulation occurs, but this is more likely to be the case for adults. Some yearlings evidently choose their nest sites after copulation has occurred. As with the typical prairie grouse, there is no evidence that a female needs to be mated more than once to obtain sufficient sperm to fertilize an entire clutch. Nest sites usually are well hidden in cover that offers vertical shading, such as sage bushes or standing grass. However, at least one or more sides should provide sufficient lateral visibility so as to assist the female in evading ground predators. There is no obvious relationship between lek location and nest location, and in five different studies involving more than 300 nests the average distance between the nest and the lek where the female was first seen or captured was 3.5 miles. This distance is greater than the mean interlek distance in several studies, which has ranged from about 0.8 to 3.0 miles. It is believed that 3 to 4 days elapse between copulation and the laying of the first egg, plenty of time to select a site and construct a nest.

Like the typical prairie grouse, females lay their eggs at the rate of somewhat less than 1 egg per day, averaging about 3 days for every 2 eggs. The average clutch size in a variety of studies has ranged from 6.6 to 9.9 eggs, but is most often between 6.5 and 7.5 eggs. It thus averages about 3 fewer eggs than are usual in the typical prairie grouse. The incubation period is longer than in prairie grouse, ranging from 25 to 27 days, a difference that might be expected given its somewhat larger egg volume.

The incidence of nesting success (percentage of nests hatching one or more chicks) in sage-grouse has been studied in at least seven states, and estimates have ranged widely, from 14.5 to 86 percent. An unweighted average from 14 field studies is about 50 percent. Anywhere from less than 10 to more than 40 percent of unsuccessful females may attempt to renest. Nesting success is related to the quality of nest cover (higher in areas rich in sage, other shrubs, forbs, and grasses; lower in areas with few shrubs, in burned or plowed areas, or in those areas that have been sprayed with shrub-killing biocides). The density of predators such as ravens and the amount of precipitation occurring during the hatching and brood-rearing season may also influence nesting and rearing success.

Estimates of annual survival rates among sage-grouse have been quite variable. It is likely that less than 50 percent of all hatched chicks will live until autumn independence. The survival of young birds of both sexes may average about 50 percent for the remainder of their first year. Survival of adult birds seems to be sex-related, with the males surviving at a generally lower rate than females. Less than half the males in a population are likely to return to their leks the following year, suggesting a surprisingly high male mortality rate. Among studies of banded or radio-tagged populations, the males seem to have survived at a rate averaging about 10–30 percent less than females. Various studies have placed the annual survival rate of yearling to adult females ranging from about 55 to 75 percent, and that of males averaging about 40 to 50 percent.

The Washington population of the western greater sage-grouse consisted of about 1,100 birds in 2000. No hunting has been allowed since 1986. In April 2001 the U.S. Fish and Wildlife Service determined that a nationally threatened status for this population was warranted but precluded. The service was required to make a legal assessment of the bird's status after two conservation groups, the Northwest Ecosystem Alliance and the Biodiversity Legal Foundation, initiated a legal petition toward that end. According to Susan Tweit, if the greater sage-grouse were to be listed as a nationally threatened species, the management of millions of acres of public-access lands would be affected, as well as some five thousand federally managed grazing allotments. Farming, hunting, surface mining, and many other activities on federal, state, and private lands would be affected over a much wider geographic scale than was impacted by the comparable listing of the spotted owl. Like the spotted owl, the sage-grouse is simply one of the most conspicuous members of a group of sage-dependent species. Its future prospects will provide a potent test of wills, placing diverse local, state, and national interest groups in varying degrees of opposition to one another. It is unlikely that the sage-grouse, sage sparrow, sage thrasher, and other sage-related biota, and indeed sagebrush as well, will come out as winners in such a battle.

GUNNISON SAGE-GROUSE

Few if any ornithologists would have believed that a wholly unknown species of large game bird could have survived into the final years of the 20th century in western North America without having at least been observed and judged to be a possible new species. Yet such is the case. Only in the 1980s when Clait Braun, a grouse specialist working for the Colorado Division of Wildlife, began examining the wings of sage-grouse killed by upland game hunters in the Gunnison Basin of southwestern Colorado did the first evidence of a possible new species surface. He noticed that the wings of these sage-grouse were considerably smaller than those of sage-grouse from elsewhere in Colorado. A marginally smaller western race of sage-grouse from the Pacific coastal states was already known, but the Gunnison birds had even smaller wing measurements. Braun investigated further, and in 1991 he and Jerry Hupp of Colorado State University jointly reported that the Gunnison Basin birds had major physical differences from grouse taken elsewhere in Colorado, not only in their body mass but also in most linear measurements, such as bill, wing, and leg lengths. Although situated only about 120 miles from the nearest population of greater sage-grouse, adult males of the Gunnison population average about 27 percent lighter in body mass, and their bill lengths average 23 percent shorter.

The most important piece of evidence relative to the uniqueness of the Gunnison birds came from an undergraduate student working on natural sounds in a lab at the University of California at San Diego. This student, Jessica Young, received some tape recordings of sage-grouse displays that had been recorded from the Gunnison Basin. On listening to them, she realized that the sounds were quite different from the usual ones made by displaying sage-grouse, whatever their subspecies. Specifically, there were far more plopping noises and wing-swishes than occur in sage-grouse elsewhere in the West.

Jessica Young and others later undertook a detailed analysis of the secondary sexual traits of Gunnison sage-grouse, specifically the sequential features of the male's strutting display. Among widely geographically distinct populations of the greater sage-grouse this complex display shows virtually no measurable variation. In an analysis of 10 acoustical display components of the Gunnison grouse, as compared with greater sage-grouse, 8 exhibited statistically significant differences (Figures 26 and 27). For example, there are 9, rather than 2, sounds associated with air sac plops in Gunnison sage-grouse. The wing-swish noises are weaker in the Gunnison birds, owing to reduced amplitudes of the

Sage-Grouse

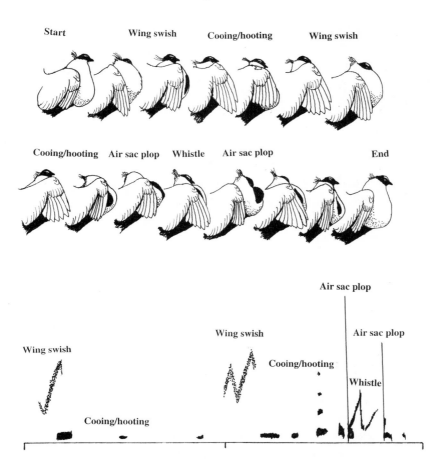

Figure 26. Greater sage-grouse, display sequence *(above)* and sonographic depiction *(below)* of strutting display by male (mainly after sketches and sonogram by Young et al. 1994); sequence duration 2.0 seconds. (Starting and ending points of display are not exactly the same as those in Figure 27.)

wing movements. The calls are all lower in frequency than in the northern species, contrary to what might have been expected in smaller birds.

The visual differences between Gunnison and greater sage-grouse are also marked. Gunnison males have thicker, longer, and more conspicuous black filoplumes that are at times raised well above the male's head during display and resemble a long, black occipital crest, or "ponytail," rather than a halolike array of feathery antennae (compare Figures 22 and 28). The dark brown back and upper wing coverts have conspicuous white shaft-streaks, but the whitish mottling on the vanes of these feathers is less than that on the highly mottled

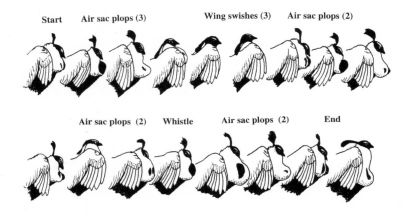
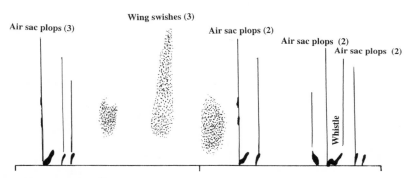

Figure 27. Gunnison sage-grouse, display sequence *(above)* and sonographic depiction *(below)* of strutting display by male (after sketches and sonogram by Young et al. 1994); sequence duration 2.0 seconds.

wing coverts of greater sage-grouse. The rectrices of both are more contrastingly patterned, with distinct cross-barring of dark brown and creamy white, rather than being rather irregularly mottled.

In both species the white feathers along the sides of the breast and in front of the wings are identically modified for mechanical sound production by having stiffened shafts that are raised during display so that the anterior wing feathers can abrasively scrape over them. The forward and backward wing movements of the Gunnison are less extreme, and there are a substantially larger number of air sac inflations and deflations within each strutting sequence, although the duration of each sequence is essentially identical, namely about two seconds (see Figures 26 and 27). The display rate (strutting sequences per minute) averaged slightly lower in the Gunnison population than in two widely separated populations of greater sage-grouse.

The ecology of the Gunnison sage-grouse is similar to that of its larger relative. The Gunnison form is closely associated with various subspecies of big sagebrush *(A. t. wyomingensis, A. t. vaseyana)* and other related sages *(A. nova* and *A. arbuscula)* throughout the year. Sage is used for escape cover, a winter food source, and thermal cover during summer. During winter the birds are prone to using areas with more deciduous shrubs and mixed pines, junipers, and oaks. Tall, dense stands of sagebrush are selected by females for nesting cover, and broods are reared in wet meadows where insects and succulent forbs are abundant, but where nearby wide stands of sagebrush are available for hiding and foraging. During winter tall stands of sagebrush are used for foraging; shorter stands on hills and ridge tops serve for roosting. Even shorter and more open stands are used as lek sites. In disturbed habitats the birds may have to rely on cultivated fields for food and cover.

In 1999 the American Ornithologists' Union recognized the Gunnison sage-grouse as a distinct species, *Centrocercus minimus*. Young, Braun, and coauthors provided a detailed description of the new species and pointed out that birds of the Gunnison type probably once occurred in but have now been extirpated from New Mexico, Oklahoma, northeastern Arizona, southwestern Kansas (its historic occurrence there is now considered doubtful), and at least nine Col-

Figure 28. Gunnison sage-grouse, male in strutting display posture with raised filoplumes (mostly after photos by Lance Beeny). Also shown are a greater primary covert *(bottom)* and lateral tail feather (from specimens, to scale; compare with Figure 22).

orado counties. They are now limited to only six or seven counties in southwestern Colorado and one adjoining county in southeastern Utah. The largest population occurs in Colorado's Gunnison and Saguache counties, which as of 2000 had a total of about 4,000 birds, most of which occur in the Gunnison Basin, largely on Bureau of Land Management lands. This population appears to have remained essentially stable through the 1990s. There are six additional small isolated populations of from 3 to 300 birds, including a Utah population of about 120 individuals. At least one of these small populations may have become extirpated, and nearly all the populations on private lands are declining. Since 1963 the average number of males attending leks in the Gunnison Basin has declined by more than 60 percent, and the surviving subpopulations are known to exhibit low genetic diversity, in a manner analogous to Africa's small and isolated cheetah populations. Genetic comparisons between the Gunnison and greater sage-grouse have revealed some small but possibly significant differences suggesting that a genetic barrier to gene flow between these populations may exist.

The World Conservation Union included the Gunnison sage-grouse as endangered in its 2001 IUCN Red List of Threatened Species. At about the same time, the American Lands Alliance and a coalition of national environmental groups petitioned the Fish and Wildlife Service to consider an emergency listing of the Gunnison sage-grouse as endangered, under the provisions of the Endangered Species Act. In January 2000 the U.S. Fish and Wildlife Service described the Gunnison sage-grouse as a listing candidate but did not support its immediate classification as either threatened or endangered. In December 2000 it was described as a candidate for listing as threatened and assigned an action priority level of 5 (out of 12 categories, with category 1 having the highest priority). Given the glacial rate of movement in the federal legal process, the birds might well be endangered before they are officially recognized as threatened and extinct before they are classified as endangered.

7

CAN THE FABRIC BE MENDED AND THE PIECES PRESERVED?

Our natural habitats of North America once resembled a beautiful tapestry in which the emerald greens of the coastal, northern, and montane forests gradually gave way to the softer summer-green and winter-golden tones of the interior grasslands, and then to the grays and browns of the deserts. Black tears and stains have now disrupted the nation's fabric, where agriculture, forestry, cities, and the other stigmata of modern life have left their marks. Of all these disruptions, none have been more devastating than those affecting the prairies, especially the midgrass and tallgrass prairies; a great gaping hole in the tapestry now exists where the Great Plains grasslands once held sway.

There, in areas once dominated by tall bluestem grasses, a substantial percentage of America's human population can now live their entire lives without ever seeing those bronzy red grasses. They may not even be vaguely aware that the occasional patches of grass that survive in rights-of-way beside the paved highways they traverse once spread as far as the eye could see in all directions and were in large part responsible for creating the rich soils on which they have built their homes, their fortunes, and their lives. Even fewer residents of the region have probably ever seen a prairie grouse.

With these thoughts in mind, the question arises whether some small parts of the prairies and their associated flora and fauna can be saved, at least in sufficient quantities to assure remnant populations for future study, for enjoyment, and to provide a degree of respect for the biodiversity that once surrounded us and still survives in a few locations. Even a relict tallgrass prairie no larger than a few hundred acres, such as the 260-acre Nine-Mile Prairie near Lincoln, Nebraska, may support nearly 400 species of native plants, to say nothing of the even larger array of animals and microorganisms living there. Most of these organisms have been studied only to the degree that they have been given scientific names. Some of the invertebrates and soil microflora have not even received that much attention. NASA has spent uncounted millions in collecting and bringing back small samples of moon soil and rocks for exhaustive scientific analysis. Yet their massive buildings near Houston were built on tallgrass prairie soils, and these facilities have completely replaced the now-vanished coastal prairie communities, which had barely been surveyed by biologists before they were completely gone.

STRATEGIES FOR PRESERVING THE PRAIRIE GROUSE

The formula for saving prairie-chickens, sharp-tailed grouse, and sage-grouse from extinction is fairly straightforward, albeit often difficult and expensive: save their habitats in sufficient size and quantity to protect their genetic diversity and overall viability.

The most conspicuous and most numerous of our large prairie mammals, the bison, was almost lost a century ago. Even today few areas of native grassland are large enough to sustain completely unconfined herds, which are now mainly limited to national parks, national monuments, and national grasslands. Only a few entrepreneurs and conservation groups, such as Ted Turner and the Nature Conservancy, have the wherewithal to provide large-scale habitat protection for bison and their associated prairie ecosystem. Grouse don't need areas as large as bison, a point in their favor, but they do require more space than almost any of the other grassland birds. By the time Nine-Mile Prairie in southeastern Nebraska was finally somewhat protected in the 1940s, it was already too small to support greater prairie-chickens, and since then it has only gotten smaller. A similar nearby tallgrass prairie (Audubon's Spring Creek Prairie) of slightly more than 600 acres is also too small, but some surrounding natural grasslands raise the total contiguous grassland area to about 1,500 acres, barely enough to support a tiny prairie-chicken flock.

PRAIRIE PRESERVES, PARKS, AND WILDLIFE REFUGES AS POPULATION REFUGIA

The estimated original regions historically occupied by tallgrass prairies has been independently estimated by Dennis Farney and David Wilcove as 231,000–400,000 square miles, out of a total native North American grassland area of about 1–1.3 million square miles. The combined mixed-grass and shortgrass regions were judged by Wilcove to encompass 625,000 square miles. For convenience, one might assume these two latter habitat types once occupied roughly comparable areas, for their common boundaries are usually indistinct and shift dynamically back and forth during wetter and drier climatic periods. This treatment would suggest that each of these three native grassland types may have once occupied about 300,000 square miles before European settlement. An alternative estimate can be obtained by basing original areas on a vegetation map prepared by A. W. Kuchler. The estimated original areas of these three mega-ecosystems are then rather smaller, the tallgrass, mixed-grass, and shortgrass components comprising about 221,000, 219,000, and 237,000 square miles, respectively. A compromise estimate for each of the three grassland types might be of about 250,000 square miles.

Many grassland preserves that support prairie grouse already exist in the United States and southern Canada, only a few of which were established primarily for maintaining grouse habitat. In my book *Prairie Birds: Fragile Splendor in the Great Plains*, I tallied all the existing major grassland preserves in the Great Plains, a region that encompasses the ranges of all the surviving forms of grouse described in this book except for the two sage-grouse and some northern populations of the prairie sharp-tailed grouse. The reserves selected for listing generally were those of at least 1,000 acres, an area just barely large enough to support prairie grouse. The tallgrass preserves collectively represented about 820 total square miles in the United States, and 31 square miles in Canada. Of all the midgrass preserves, there were 1,200 square miles in the United States and 1,300 square miles in Canada. Of the shortgrass preserves, generally too dry for farming, there were 75,500 square miles in the United States and almost none in Canada. These total preserved areas sometimes include extensive non-grassland habitats, such as wetlands and variably wooded areas. Nonetheless, the figures provide a fairly accurate estimate of the maximum preserves.

Comparing the previously estimated historical grassland areas (roughly 250,000 square miles each for tallgrass, mixed-grass, and shortgrass prairies) with the collective areas of the preserved grassland sites, one might tentatively conclude that about 0.3 percent of the original tallgrass prairie is now protected from further destruction, as well as about 1.0 percent of the midgrass prairies, and about 30 percent of the shortgrass prairies. It can be no surprise that the tallgrass-adapted greater prairie-chicken has suffered the most in historic times, and the much more broadly distributed sharp-tailed grouse, the least.

As for the sage-grouse, the estimated historical range of its close ecological associate big sagebrush was approximately 150,000 square miles. Several national parks and national monuments or national grasslands are located within the range of big sagebrush, which still occupies perhaps half its original range. However, only a few of these preserves also support populations of greater sage-grouse, and the total range and fate of the closely related Gunnison sage-grouse is now largely under the control of the Bureau of Land Management (BLM).

Of the federal agencies charged with protecting our natural heritage, the BLM administers the greatest area, some 526,000 square miles in nearly 50 separate administrative districts. These lands are mostly in regions too arid to support agriculture, but they provide wonderful habitat for sage-grouse and related sage-steppe biota. They are also coveted by ranchers, long accustomed to grazing BLM lands at cut-rate prices. Their cattle and sheep effectively trample the soil, compacting it and increasing erosion rates, as well as reducing natural cover and increasing the invasion rates of species such as various undesirable nonnative weeds and conifers. In the last few years the BLM has begun shifting its emphasis away from satisfying livestock and oil-and-gas interests toward active protection and management of wildlife habitat, a move long overdue and greatly to be applauded.

The National Forest Service (NFS) has the next largest national domain. It controls nearly 300,000 square miles of more than 140 national forests, about 20 national grasslands, and numerous scenic trails, some of which are also prime habitat for plains sharp-tailed grouse and sage-grouse. It has long been the favorite friend of the logging industry, with wildlife conservation far down on its list of priorities, although a few of its national grasslands probably offer the last, best hope of preserving lesser prairie-chickens. As with the BLM, an increased awareness at the NFS of its importance in preserving prairie wildlife, especially on its national grasslands, has developed in recent years.

The National Park Service controls about 125,000 square miles of prime natural habitats, with more than 320 designated nationally protected sites, including more than 100 national parks or national monuments, plus additional seashores, lakeshores, rivers, historic sites, recreation areas, and other national preserves. Several of the national parks and monuments contain large areas of prairies, especially shortgrass prairie. Luckily, its lands are inviolate to nearly all destructive uses, and the nation's many national parks and monuments are major refugia for the protection of rare animal and plant species.

The U.S. Fish and Wildlife Service controls about 145,000 square miles of prime wildlife habitat, including nearly 400 national wildlife refuges. In a summary of bird checklists from 210 nature preserves, mostly comprised of national wildlife refuges, John O. Jones has tallied 20 preserves supporting populations of sharp-tailed grouse, 16 having populations of sage-grouse, 12 with greater prairie-chickens, and 3 with lesser prairie-chickens. In a similar but more geographically restricted tally of the Great Plains sanctuaries and prairie preserves I listed in *Prairie Birds,* 15 preserves supported sharp-tailed grouse, at least 6 and probably 9 had greater prairie-chickens (4 North Dakota prairie refuges that probably all support prairie-chickens were grouped in a single list), and 2 had lesser prairie-chickens. Sage-grouse were not included in this survey, which was limited to the Great Plains endemic bird species.

It is clear that fairly large areas of public lands such as those just mentioned now provide substantial habitats for sharp-tailed grouse and sage-grouse; fewer exist for greater prairie-chickens; and only a very few offer any secure habitat for lesser prairie-chickens. These latter include the contiguous Cimarron and Comanche National Grasslands (total 530,000 acres) in Kansas and Colorado, respectively, the Optima (4,300 acres) and Washita (8,200 acres) National Wildlife Refuges in Oklahoma, and Muleshoe National Wildlife Refuge (5,800 acres) in Texas. But the birds are now rare in all these refuges, and at the Cimarron and Comanche National Grasslands they are uncommon at best.

BLM lands in New Mexico that support shinnery oak habitat are extensive, consisting of perhaps as much as 1.2 million acres. Yet the lesser prairie-chicken population in the state verges on endangered, as the BLM land is primarily managed for cattle grazing and big game hunting, as well as facilitating gas and oil exploitation. The Black Kettle National Grassland (31,000 acres) in Roger Mills County, Oklahoma, and the state-owned Packsaddle Wildlife Management Area (16,000 acres), in Ellis County, Oklahoma, represent practically the only publicly owned shinnery oak habitat suitable for lesser prairie-chickens outside of New Mexico. Of these, only the latter location might actually have good prairie-chicken populations but as far as I know they are still unmeasured. The Black Kettle grasslands have been too fragmented by the effects of past agricultural disturbances to support prairie-chickens in even small numbers, and a biologist stationed there told me he has never even seen one on Black Kettle lands.

The undisturbed shinnery grassland ecosystem generally supports even higher densities of lesser prairie-chickens than do the sandsage grasslands, as well as good populations of scaled quail, Chihuahuan ravens, loggerhead shrikes, and black-throated sparrows, all of which are seriously declining species at the national level. Yet as Roger Peterson and Chad Boyd noted, most of the research that has been directed toward the shinnery oak ecosystem has been devoted to its eradication, mainly because it may harbor overwintering boll weevils and its leaves are seasonally toxic to cattle. Treating shinnery with defoliant herbicides has been found to sharply decrease prairie-chicken and scaled quail populations. Effects on other breeding bird species are varied, with some open grassland species such as meadowlarks and shrikes substantially increasing. Defoliation treatments may also reduce the endemic and state-endangered (in New Mexico) sand dune lizard *(Sceloperus arenicolus)* population as much as sixfold.

Clearly a new national wildlife refuge, or comparable nature preserve, is needed for the lesser prairie-chicken. Such a preserve could well be located in the now unprotected sandsage grasslands of the Arkansas River Valley from

Garden City, Kansas, west to at least the Colorado border, perhaps the best of the species' remaining range and one not yet seriously affected by cattle overgrazing. Such a preserve would offer the best hope for saving the lesser prairie-chicken from the disastrous recent history of the Attwater's prairie-chicken, when the federal government delayed far too long before starting to acquire critical habitat for its preservation. Protecting the lesser prairie-chicken there would also help protect the rapidly declining national populations of lark, grasshopper, and Cassin's sparrows, burrowing owls, and black-tailed prairie-dogs. The valley is also an extremely important migratory stopover site for several other prairie endemics, the Baird's sparrow, lark bunting, long-billed curlew, and McCown's longspur. It encapsulates the entire sandsage ecosystem, one of the rarest and least studied of the Great Plains vegetational complexes.

MINIMUM HABITAT REQUIREMENTS AND MINIMUM VIABLE POPULATION SIZES

Deciding "how much is enough?" is a problem that everyone faces at times. It is a question increasingly faced by conservation agency administrators, who must frequently decide whether a piece of land is worth trying to save at all, and if so, what is the least area that is acceptable given the constraints of time, resources, and energies needed to achieve as many of the desired goals as possible. It might be a fairly easy choice when the objective is to save a local population of some rare plant whose actual and potential habitat limits may be fairly readily evaluated. It is much more difficult with mobile species, especially migratory ones, where controlling the species' overall yearly habitat needs may be impossible. With prairie grouse the answer probably lies somewhere between these extremes. If not truly migratory, prairie grouse are surprisingly mobile. Seasonal movements of 10 to 20 miles by sharp-tailed grouse and prairie-chickens are not unusual; and annual movements of 25 to 50 miles are known. Those of sage-grouse may easily be twice as great, and annual movements of up to 150 miles may occur.

Thus buffer zones extending well beyond the usual home range of a single viable flock are desirable. Ronald Westemeier and Sharon Gough suggested that for greater prairie-chickens, minimum viable populations of at least 100 males, and preferably more than 250, are desirable, located in areas where demographic and genetic exchange are possible. They also suggested that minimum areas of suitable grasslands for supporting such populations may range from 1,500 to 13,000 acres in various parts of the species' range. An estimate of 4,000 acres of suitable grassland was made by J. Toepfer and others in order to sustain a population of 200 to 250 male greater prairie-chickens in Minnesota and Wisconsin, which might serve as a general comprise estimate for that species and possibly also sharp-tailed grouse. Sage-grouse would certainly need larger areas, but lesser prairie-chickens perhaps less.

In regard to minimum preserve sizes, it might be remembered that the area preserved for protecting the heath hen on Martha's Vineyard was less than 2,000 acres. This area proved to be far too small and resulted in the compression of the population in such a way as to make it highly vulnerable to local outbreaks of fire and disease. Although the Attwater Prairie Chicken National Wildlife Refuge in Texas is larger (8,000 acres), it too has proven too small as well as ecologically fragmented and degraded. The overall remaining Attwater's populations are now too isolated and tiny to offer any hope for the birds' long-tem survival.

Both these well-documented prairie-chicken populations suffered similar and rather rapid population declines once their populations dropped below a few thousand birds. The vagaries of weather, the resultant uncertainties of reproductive success from year to year, and other often unpredictable mortality factors have meant that more and more of a rare grouse's chances for survival rest on the reproductive fortunes and survival of progressively fewer birds. These problems are exacerbated by the lek breeding system, which works only when populations are dense enough to provide the visual and acoustic stimuli necessary to attract prebreeding females to a functioning lek, and when the lek is large enough to assure that the matings will be performed by the genetically and physically fittest males. Lek breeding also tends to promote inbreeding owing to the limitations caused by only a few males siring offspring. Over prolonged periods this trend can produce "genetic bottlenecks," which seemingly occurred in the Illinois flock of greater prairie-chickens as it spiraled downward. It rebounded once some new blood was introduced by releasing birds from states farther west into the Illinois population.

THE ROLE OF FIRE IN GROUSE MANAGEMENT

Fire is one of those things that must be handled with great care: too much, too little, too early, too late, too hot, and too cool are all variables that conservationists must take into account when contemplating the use of prescribed fire in habitat management for a particular species, to say nothing of the highly varied, if not opposite, effects the same fire might have on different but equally desirable species. The effects of fire are often apparent almost immediately, although it may take years or decades for a piece of land to recover from the undesirable effects of a single unplanned or badly handled fire,

As an example of the complexities involved in fire management, some species of sagebrush such as sandsage regularly resprout following fire and thus may actually benefit from periodic fires. More fire-sensitive sage species such as big sagebrush and its near relatives are readily killed by exposure to fire. Up to 30 years may be needed for stands of big sagebrush to recover from a single fire. Thus although fires may kill the sagebrush that sage-grouse need for winter foods and nesting, the forbs that are stimulated following fires may be selectively chosen for summer foraging by the birds. Fires occurring at long intervals (up to about 50 years) may also help keep junipers and piñon pines from invading sagebrush; the presence of these conifers in sagebrush habitats seemingly attracts predators and thus reduces their use by sage-grouse. In the sand-sage shinnery community, fires generally stimulate regrowth by both sandsage and shinnery oak, the latter reaching 3 to 4 feet in height within a few years following fire. As with other mixed shrub and grass communities, the productivity of perennial grasses may also be stimulated by periodic burns, thus improving nesting or brooding cover. *Artemisia* species are generally long-lived plants, with some big sagebrush reportedly surviving for a century or more. Yet it cannot resprout after burning; it is sensitive to flooding; and it can be reestablished only from seeds. With the invasion of annual grasses such as cheatgrass, fires are more frequent and occur earlier than was true in the past, and several less desirable shrub genera *(Chrysothamnus, Ephedra,* and *Tetradymia)* that are able to resprout following fires may gradually replace big sagebrush following an episode of fire. Fire at any season is apparently detrimental to sage-grouse in Idaho, judging from observations by J. W. Connelly, K. P. Reese, and others.

In general, controlled burns for prairie-chickens and other prairie grouse should be done a full month before nest initiation, and burns should be performed in a mosaic pattern over a several-year period. Grazing or cutting of burned grasslands should be moderate, and cutting should be done only once per season, after the early brooding season has been completed. Probably no more than half the new growth should be cut for forage, and a small part of each pasture should be left entirely ungrazed and unharvested. Lek sites need to be kept at a low vegetational stature, but nearby grassy areas should be kept tall enough to provide nesting cover.

In more typical grasslands, as in tallgrass prairies, burns likewise typically tend to favor grasses over shrubs. Depending on the severity of the fires, invading shrubs and small saplings may thus be killed or at least controlled, whereas older trees might readily survive occasional ground fires. If the fires are timed properly, warm-season grasses are also favored over the generally less desirable cool-season forms, such as various annual brome grasses that usually have limited value to prairie grouse as cover or food.

Although spring burning usually improves the stands of native, mostly warm-season, grasses, but side-oats grama seems to be harmed by frequent spring fires. Fires later in the growing season tend to reduce both warm-season and cool-season grasses. Winter burns are prone to favor cool-season grasses over the generally more desirable warm-season species. For example, an April burn in Nebraska tallgrass prairie provides a charred litter layer on the soil that not only immediately releases valuable nutrients but also helps warm the upper soil layer as sunlight directly warms the soil surface. This in turn favors an early growth spurt by warm-season grasses such as the bluestems, to the disadvantage of cool-season forms such as the bromes and especially to shrubs such as sumacs and invading junipers. However, too-frequent fires will reduce the diversity of legumes and other forbs in prairie communities, many of which provide seeds or leafy parts rich in proteins. They may also reduce the litter layer to the point that some endemic prairie birds such as Henslow's sparrows, which nest in well-developed surface litter, will cease to breed in the area. Further, native warm-season grasses that are left unmanaged and completely unmowed may not provide suitable nesting or brood-rearing habitat for prairie grouse, or may not offer any low-cover sites suitable for lek locations. Planning and conducting prescribed fires is in many ways a case of trying to play God and hoping that the decisions made are going to provide the generally short-term effects hoped for and not have too many unplanned and undesirable long-term effects.

AGRICULTURE AND GRAZING EFFECTS

Ironically small-grain agriculture was primarily responsible for producing the remarkable population explosion and range expansion of the greater prairie-chicken into the central and western plains during the late 1800s and early 1900s, only to cause its decline and near-disappearance toward the end of the 20th century. Like many things in nature, a small degree of change in the form of a new source of fall and winter food was highly beneficial; only when other seasonal habitat needs began to decline and finally disappear did the balance swing in the other direction. As agriculture became more mechanized and the farms larger, fewer and fewer acres were left as weedy edges or spared from increasingly effective pesticides and herbicides, and the prairie grouse began a long, slow, but certain descent into oblivion.

The almost universal declines in prairie grouse populations can be attributed largely to agricultural or grazing influences on habitat quality and quantity. Neither predators, disease, parasites, competitors, nor hunting has had anything approaching the degree of undesirable influence brought about by agriculture and grazing. The federal government's Conservation Reserve Program, in which farmers have been encouraged to plant highly erodable land to grasses, specifically (in recent years at least) native grasses, is one of the few examples of beneficial agricultural practices. This program has been responsible, at least in part, for regional increases in the prairie grouse populations of western Nebraska and eastern Colorado, and perhaps elsewhere. These grasses may need occasional mowing and prescribed burning to provide optimum wintering and nesting cover.

With heavy cattle grazing of tallgrass prairie, several of the major native prairie plants typical of prairie grouse breeding habitats respond as "decreasers," including little bluestem, big bluestem, Indian grass, and switchgrass, especially the last two. Needle-and-thread, side-oats, and blue grama are typical "increasers," and under heavy grazing, buffalo grass may eventually replace most of the higher-stature grasses. Annual alien grasses such as several bromes, especially cheatgrass (downy brome), become serious and pernicious invaders with overgrazing. Additionally, litter cover declines with overgrazing, increasing runoff and decreasing water infiltration. All these changes are likely to have adverse effects on prairie grouse. Among mixed-grass prairies similar effects occur during grazing, but blue grama, sand dropseed, and paspalum grasses are increasers. Effects of drought on the mixed-grass prairies are nearly the same as those of severe overgrazing, converting mixed-grass communities to shortgrass ones and making them essentially unsuited to plains sharp-tailed grouse.

ECOTOURISM AND THE FUTURE OF PRAIRIE GROUSE

Like fire and wilderness, ecotourism must be treated with great care. Aldo Leopold once wrote that the only way we can enjoy wilderness personally is by visiting it, and in so doing, we help destroy some of the very attributes of wilderness we most appreciate.

Prairie grouse must likewise be handled with great care if they are to be enjoyed in nonconsumptive ways. They are an increasingly rare commodity, and like many other attractive but rare types of wildlife can sometimes be loved almost to death. Some of the richest moments of my life have been spent in grouse blinds, and I fervently believe that a naturalist has never fully lived unless he or she has watched the sun rise on a grouse lek. Yet it is not always a harmless pleasure.

For more than 40 years I have watched such sunrises on a hillside in southeastern Nebraska near Burchard Lake, where each spring morning, perhaps for centuries, prairie-chickens have played out their roles of survival and reproduction. In the early years of the 1960s, more than 40 males could reliably be seen each spring, and the lek was visited by only a few people. But during the 1970s the state's Game and Fish Commission built two large blinds on edges of the leks, blinds that were badly designed, with too-large and noncamouflaged windows, and no screening for concealing approaches or departures. These blinds increasingly attracted more and more visitors, many of whom knew little or nothing of the proper etiquette of grouse watching, especially the need to arrive at dark and to stay until the birds had finished their daily routines. Over the years, fewer and fewer males appeared, and those that did became increasingly wary of the blinds. By the late 1990s only about a dozen males regularly used the lek. In 2001, after an experimental hunting season in southeastern Nebraska, only four males appeared. Four males are too few to attract females effectively, and it is likely that this entire lek will eventually disappear.

A somewhat similar downward population progression occurred on the 8,600-acre Konza Prairie in the Flint Hills of Kansas, where greater prairie-chickens were once quite common. From 1980 to 1990 the total number of males n this locally protected population declined 68 percent, and the number of active leks was reduced by 38 percent, while during the same period grouse research projects increased threefold. Meantime, prairie-chicken populations on the nearby grasslands remained stable. The Konza Prairie population of prairie-chickens is still quite low. We can hope that the value of the research gained at Konza and the personal enjoyment and love for prairie wildlife gained by the bird-watchers who have visited Burchard Lake over the past four decades are worth such costs. This may well be the case, for many of the people I have taken to watch the prairie-chickens at Burchard Lake have since become outspoken advocates for the preservation of prairies and prairie wildlife, and one woman told me decades later that visiting it as an ornithology student had provided the most exciting and personally rewarding experience of her entire life. A nonbiologist, indeed a practicing Protestant minister, told me it had been his most religious experience. What might happen to a person when first visiting a grouse lek is impossible to predict, but it is likely to be an unforgettable event.

We cannot foretell how many people's lives might be affected or even fundamentally changed through the magic of observing wild prairie grouse in their natural habitat as they engage in the calls, postures, and behaviors shaped by millions of years of evolution and try their best to pass them on to the next generation. There are promises to be witnessed and given thus: promises on our part to help save the land and its living beings for future generations, and promises on the part of the birds that, so long as they might live, they will try to fulfill their end of the bargain. I can think of none better.

IDENTIFICATION KEY TO THE NORTH AMERICAN PRAIRIE GROUSE

A. Tail feathers variably pointed at tips and the tail graduated in shape; the middle feathers considerably longer than the outermost pair
   1. Tail shape distinctly pointed; 18 feathers, the middle pair more than twice the length of the shortest, and the most heavily patterned . . . **Sharp-tailed grouse** *(Tympanuchus phasianellus)*
   2. Tail shape less strongly graduated; all 20 feathers sharply pointed and barred with buff, brown, and fuscous
      a. Larger, the male with wispy filoplumes on the sides of the neck; the tail feathers (rectrices) irregularly patterned in both sexes; the darker areas much broader than the light ones . . . **Greater sage-grouse** *(Centrocercus urophasianus)*
      b. Smaller, the adult male with broader-tipped filoplumes on sides of neck and the rectrices more clearly cross-barred; the light bars almost as broad as the darker ones . . . **Gunnison sage-grouse** *(Centrocercus minimus)*
B. Tail feathers rounded at tips and the overall tail shape also rounded; neck feathers with ornamental pinnae, especially in males ("Pinnated grouse")
C. Darker bars of back and rump divided, with a brown bar enclosed between two narrow blackish ones; breast feathers with 4–6 narrow bars of brown and white; side and flank feathers bicolored, with two narrow blackish bars enclosing a broader pale brown one; pinnae of adult male relatively long (ave. ca. 1.5–1.6 × tarsus length); air sacs dull red. Adult

males ave. 730–815 g; females 630–740 g . . . **Lesser prairie-chicken** *(Tympanuchus pallidicinctus)*

1. Darker bars of back and rump solid blackish brown; breast feathers with single brown bar and whitish tips; side and breast feathers barred broadly with brown; pinnae of adult males relatively short (ave. ca. 1.2–1.4 × tarsus length); air sacs orange to orange yellow. Adult males ave. 940–1,040 g; females 730–890 g . . . **Greater prairie-chicken** *(Tympanuchus cupido)*
    a. Pinnae of adult male with up to about 12 elongated and rather pointed feathers; tawny to cinnamon rufous above, darker below; the olive brown to fuscous barring on underparts rather broad; axillaries banded or heavily spotted, scapulars with conspicuous whitish terminal spots . . . **Heath hen** *(Tympanuchus cupido cupido)* **(extinct)**
    b. Pinnae of adult male with more than 12 squarish tipped feathers; cinnamon to tawny olive above, paler buffy below, the underpart barring narrower; axillaries usually pure white; scapulars usually without whitish terminal spots
        aa. Lower legs (tarsi) feathered to base of toes; longest pinnae ave. 65 mm (range 50–83 mm); more tawny olive on head and neck . . . **Interior greater prairie-chicken** *(Tympanuchus cupido pinnatus)*
        bb. Tarsi unfeathered along rear and on lower parts of front edge, especially in summer; longest pinnae ave. 53 mm (range 46–73 mm); more cinnamon on head and neck . . . **Attwater's greater prairie-chicken** *(Tympanuchus cupido attwateri)*

BIBLIOGRAPHY

Able, K. P. 2000. Gleanings from the technical literature: Sage grouse futures. *Birding* 32:306–316.

Ahlborn, G. G. 1980. Brood-rearing habitat and fall–winter movements of lesser prairie chickens in eastern New Mexico. Master's thesis, New Mexico State Univ., Las Cruces.

Aldrich, J. W. 1963. Geographic orientation of American Tetraonidae. *J. Wildl. Manage.* 27:529–545.

Aldridge, C. L. 2000. Reproduction and habitat use by sage grouse in a northern fringe population. Master's thesis, Univ. of Regina, Regina, Saskatchewan.

Aldridge, C. L., and R. M. Brigham. 2001. Nesting and reproductive activities of greater sage-grouse in a declining northern fringe population. *Condor* 105:537–543.

Aldridge, C. L., S. J. Oyler-McChance, and R. M. Brigham. 2001. Occurrence of greater sage-grouse × sharp-tailed grouse hybrids in Alberta. *Condor* 103:657–660.

American Ornithologists' Union. 1998. *Check-list of North American birds.* 7th ed. American Ornithologists' Union, Washington, D.C.

———. 2000. Forty-second supplement to the American Ornithologists' Union Check-list of North American Birds. *Auk* 117:847–858. (Gunnison sage-grouse recognized.)

Ammann, G. A. 1957. The prairie grouse of Michigan. Michigan State Dept. of Conserv. Tech. Bulletin.

———. 1963. Status and management of sharp-tailed grouse in Michigan. *J. Wildl. Manage.* 27:802–809.

Anderson, R. K. 1970. Orientation in prairie chickens. *Auk* 88:286–290.

Apa, A. D. 2000. Habitat use and movements of sympatric sage and sharp-tailed grouse in southeastern Idaho. Ph.D. diss., Univ. of Idaho, Moscow.

Applegate, R. D. 1993. Did the heath hen occur in Maine? *Maine Nat.* 1 (2): 1–4.
———. 1997. Introductions of sharp-tailed grouse into Maine. *Northeastern Nat.* 4 (2): 105–110.
Askins, R. A. 2000. *Restoring North America's birds: Lessons from landscape ecology.* Yale Univ. Press, New Haven, Conn.
Bagg, A. C., and S. A. Eliot Jr. 1937. *Birds of the Connecticut Valley in Massachusetts.* Hampshire Bookshop, Northampton, Mass.
Bailey, F. M. 1928. *Birds of New Mexico.* New Mexico Dept. of Fish and Game, Santa Fe.
Bailey, J. A., J. Klingel, and C. A. Davis. 2001. Status of nesting habitat for lesser prairie-chicken in New Mexico. *Prairie Nat.* 32:149–156.
Bailey, J. A., and S. O. Williams III. 2000. Status of the lesser prairie-chicken in New Mexico, 1999. *Prairie Nat.* 32:157–168.
Bailey, V. 1905. *Biological survey of Texas.* U.S. Biol. Survey., No. Amer. Fauna 25:1–222. Washington, D.C.
Baker, M. F. 1953. Prairie chickens of Kansas. Univ. of Kansas Mus. Nat. Hist. and Biol. Survey of Kansas, Misc. Pub. 5.
Ballard, W. B., Jr., and R. J. Robel. 1974. Reproductive importance of dominant male greater prairie-chickens. *Auk* 91:75–85.
Baumgartner, F. A., and A. M. Baumgartner. 1992. *Oklahoma bird life.* Univ. of Oklahoma Press, Norman.
Baydack, R. K. 1988. Characteristics of sharp-tailed grouse, *Tympanuchus phasianellus,* leks in the parklands of Manitoba. *Can. Field-Nat.* 102:39–44.
Baydack, R. K., and D. A. Hein. 1987. Tolerance of sharp-tailed grouse to lek disturbance. *Wildl. Soc. Bull.* 15:535–539.
Beetle, A. A. 1960. A study of sagebrush, the section *Tridentatae* of *Artemisia.* Univ. of Wyoming Agric. Exper. Station Bulletin 368:1–83.
Bent, A. C. 1932. *Life histories of North American gallinaceous birds: Orders Galliformes and Columbiformes.* U.S. Govt. Printing Office, Washington, D.C.
Berg, W. E. 1990. Sharp-tailed grouse management problems in the Great Lakes states: Does the sharp-tail have a future? *Loon* 62:42–45.
———. 1997. The sharp-tailed grouse in Minnesota. *Minnesota Wildlife Report* 10:1–17.
———. 2001. Minnesota grouse, 2001. Sutton Avian Res. Center, Bartlesville, Okla. *Prairie Grouse Tech. Newsletter* 3:12–13.
Berger, R. P., and R. K. Baydack. 1992. Effects of aspen succession on sharp-tailed grouse, *Tympanuchus phasianellus,* in the Interlake region of Manitoba. *Can. Field-Nat.* 106:185–191.
Bergerud, A. T., and M. W. Gratson, eds. 1988. *Adaptive strategies and population ecology of northern grouse.* Univ. of Minnesota Press, Minneapolis.
Bidwell, T. G., C. B. Green, A. D. Peoples, and R. E. Masters. 1995. Prairie chicken management in Oklahoma. Oklahoma Cooperative Extension Service Circular E-945.
Bjugstad, A. J., ed. 1988. Prairie-chickens on the Sheyenne National Grasslands. U.S. Dept. Agric., Forest Serv., Gen. Tech. Rep. RM-159.

Bibliography

Boisvert, J. H., K. P. Rose, and R. W. Hoffman. 2000. What Columbian sharp-tailed grouse are saying about Conservation Reserve Program and Post-Act Mine Reclamation lands in northwestern Colorado. Paper presented at 22nd Western States Sage and Columbian Sharp-Tailed Grouse Symposium, Redmond, Oreg., July 13–14, 2000. Abstract.

Bosquet, K. R., and J. J. Rotelle. 1998. Reproductive success of sharp-tailed grouse in Montana. *Prairie Nat.* 30:63–70.

Bowen, D. E., R. J. Robel, and P. G. Watt. 1976. Habitat and investigators influence artificial ground nest losses: Kansas. *Kansas Acad. Sci.* 79:141–147.

Bowman, T. J., and R. J. Robel. 1977. Brood break-up, dispersal, mobility, and mortality of juvenile prairie chickens. *J. Wildl. Manage.* 41:27–34.

Braun, C. E. 1995. Distribution and status of sage grouse in Colorado. *Prairie Nat.* 27:1–9.

———. 1998. Sage grouse declines in western North America: What are the problems? *Proceedings of the Western Assoc. State Fish and Wildl. Agencies* 78:139–156.

Braun, C. E., R. B. Davies, J. R. Dennis, K. A. Green, and J. L. Sheppard. 1992. *Plains sharp-tailed grouse recovery plan.* Colorado Dept. of Natural Resources, Div. of Wildlife, Denver.

Braun, C. E., K. W. Harmon, J. A. Jackson, and C. D. Littlefield. 1978. Management of national wildlife refuges in the United States: Its impacts on birds. *Wilson Bull.* 90:309–321.

Braun, E. L. 1964. *Deciduous forests of eastern North America.* Hafner Pub., New York.

Brewster, W. 1885. The heath hen of Massachusetts. *Auk* 2:80–84.

Brush, G. S., C. Lenk, and J. Smith. 1980. The natural forests of Maryland. *Ecol. Monogr.* 50:77–92.

Busby, W. H., and J. L. Zimmerman. 2001. *Kansas breeding bird atlas.* Univ. Press of Kansas, Lawrence.

Caldwell, P. J. 1976. Energetic and population consideration of sharp-tailed grouse in the aspen parklands of Canada. Ph.D. diss., Kansas State Univ., Manhattan.

Campbell, H. 1950. Note on the behavior of marsh hawks towards lesser prairie chickens. *J. Wildl. Manage.* 14:477–478.

———. 1972. A population study of lesser prairie chickens in New Mexico. *J. Wildl. Manage.* 36:689–699.

Candelaria, M. A. 1979. Movements and habitat-use by lesser prairie chickens in eastern New Mexico. Master's thesis, New Mexico State Univ., Las Cruces.

Cannon, R. W. 1980. Current status and approaches to monitoring populations and status of lesser prairie chickens in Oklahoma. Master's thesis, Oklahoma State Univ., Stillwater.

Cannon, R. W., and D. M. Christisen. 1984. Breeding range and population status of the greater prairie-chicken in Missouri. *Transactions of the Missouri Acad. Sci.* 18:33–39.

Cannon, R. W., and F. L. Knopf. 1979. Lesser prairie-chicken responses to range fires at the booming ground. *Wildl. Soc. Bull.* 7:44–46.

———. 1980. Distribution and status of the lesser prairie chicken in Oklahoma.

Pp. 71–74 in *Proceedings of the Prairie Grouse Symposium*, P. A. Vohs Jr. and F. L. Knopf, eds. Oklahoma State Univ., Stillwater.

———. 1981a. Lek numbers as a trend index to prairie grouse populations. *J. Wildl. Manage.* 45:776–778.

———. 1981b. Lesser prairie chicken densities on shinnery oak and sand sagebrush rangelands in Oklahoma. *J. Wildl. Manage.* 45:521–524.

Cartright, K. S. 2000. Influence of landscape, land use, and lek attendance of greater prairie-chicken lek surveys. Master's thesis, Kansas State Univ., Manhattan.

Cogar, V. F. 1980. Food habits of Attwater's prairie-chicken in Refugio County, Texas. Ph.D. diss., Texas A&M Univ., College Station.

Commons, M. L. 1997. Movement and habitat use by Gunnison sage grouse *(Centrocercus minimus)* in southwestern Colorado. Master's thesis. Univ. of Manitoba, Winnipeg.

Conard, H. S. 1935. The plant associations of central Long Island. *Amer. Midl. Nat.* 16:433–516.

Connelly, J. W., M. W. Gratson, and K. P. Reese. 1998. Sharp-tailed grouse *(Tympanuchus phasianellus)*. No. 354 in *The birds of North America*, A. Poole and F. Gill, eds. Academy of Natural Sciences, Philadelphia, and American Ornithologists' Union, Washington, D.C.

Connelly, J. W., K. P. Reese, R. A. Fisher, and W. L. Wakkinen. 2000. Response of a sage grouse breeding population to fire in southeastern Idaho. *Wildl. Soc. Bull.* 28:90–96.

Copelin, F. F. 1963. The lesser prairie chicken in Oklahoma. Oklahoma Dept. of Wildlife Conservation, Tech. Bulletin 6.

Crawford, J. A. 1978. Morphology and behavior of greater × lesser prairie-chicken hybrids. *Southwestern Nat.* 23:591–596.

———. 1980. Status, problems, and research needs of the lesser prairie chicken. Pp. 1–7 in *Proceedings of the Prairie Grouse Symposium*, P. A. Vohs Jr. and F. L. Knopf, eds. Oklahoma State Univ., Stillwater.

Crawford, J. A., and E. G. Bolen. 1973. Spring use of stock ponds by lesser prairie chickens. *Wilson. Bull.* 85:471–472.

———. 1975. Spring lek activity of the lesser prairie chicken in west Texas. *Auk* 92:808–810.

———. 1976a. Effects of land use on lesser prairie chickens in Texas. *J. Wildl. Manage.* 40:96–104.

———. 1976b. Effects of lek disturbances on lesser prairie chickens. *Southwestern Nat.* 21:238–240.

———. 1976c. Fall diet of lesser prairie chickens in west Texas. *Condor* 78:142–144.

Christisen, D. M. 1969. National status and management of the greater prairie-chicken. *Transactions of the No. Amer. Wildl. and Nat. Res. Conf.* 34:207–217.

Davis, C. A., T. Z. Riley, R. A. Smith, H. R. Suminski, and D. M. Wisdom. 1979. Habitat evaluation of lesser prairie chickens in eastern Chaves County, New Mexico. Final Report to Bureau of Land Management, Roswell, N.M. Contract YA-512-CT6-61.

Day, G. M. 1953. The Indian as an ecological factor in the northeastern forest. *Ecology* 34:329–346.

Dickerman, R. W., and J. P. Hubbard. 1994. An extinct subspecies of sharp-tailed grouse from New Mexico. *Western Birds* 25:128–136.

Dimare, M. I. 1991. Effects of lek shape on reproductive behavior of Attwater's prairie chicken *(Tympanuchus cupido attwateri)*. Ph.D. diss., Texas A&M Univ., College Station.

Dinsmore, J. J. 1994. *A country full of game.* Univ. of Iowa Press, Iowa City.

Doerr, T. B., and F. S. Guthery. 1980. Effects of shinnery oak control on lesser prairie chicken habitat. Pp. 59–63 in *Proceedings of the Prairie Grouse Symposium,* P. A. Vohs Jr. and F. L. Knopf, eds. Oklahoma State Univ., Stillwater.

Donaldson, D. D. 1969. Effect on lesser prairie chickens of brush control in western Oklahoma. Ph.D. diss., Oklahoma State Univ., Stillwater.

Drut, M. S. 1994. *Status of sage grouse with emphasis on populations in Oregon and Washington.* Audubon Society of Portland, Portland, Oreg.

Ducey, J. E. 2000. *Birds of the untamed West: The history of birdlife in Nebraska, 1750 to 1875.* Making History, Omaha, Nebr.

Ehrlich, P. R., D. S. Dobkin, and D. Wheye. 1992. *Birds in jeopardy: The imperiled and extinct birds of the United States and Canada.* Stanford Univ. Press, Stanford, Calif.

Ellsworth, D. L., R. L. Honeycutt, N. J. Silvy, K. D. Rittenhouse, and M. H. Smith. 1994. Mitochondrial-DNA and nuclear-gene differentiation in North American prairie grouse (genus *Tympanuchus*). *Auk* 111:661–671.

Evans, R. M. 1961. Courtship and mating behavior of sharp-tailed grouse. Master's thesis, Univ. of Alberta, Edmonton.

Farney, D. 1980. The tallgrass prairie: Can it be saved? *Natl. Geog. Mag.* (Jan.): 37–61.

Field, G. W. 1907. The heath hen. *Bird-Lore* 9:249–255.

———. 1913. The present status of the heath hen. *Bird-Lore* 15:352–358.

Finch, D. M. 1992. Threatened, endangered, and vulnerable species of terrestrial vertebrates in the Rocky Mountain Region. U.S. Dept. of Agric., Forest Serv., Gen. Tech. Rep. RM–215.

Forbush, E. H. 1916. *A history of the game birds, wild-fowl, and shore birds of Massachusetts and adjacent states.* 2nd ed. Massachusetts State Board of Agriculture, Boston.

———. 1918. The heath hen on Martha's Vineyard. *Amer. Mus. J.* 18:278–285.

———. 1927. *Birds of Massachusetts and other New England states.* Pt. 1. *Land birds from bob-whites to grackles.* Massachusetts Dept. of Agriculture, Boston.

Forman, R. T., ed. 1979. *Pine barrens: Ecosystem and landscape.* Academic Press, New York.

Foster, D. R. 1992. Land-use history (1730–1990) and vegetation dynamics in central New England, U.S.A. *J. Ecol.* 80:753–772.

———. 1999. *Thoreau's country: Journey through a transformed landscape.* Harvard Univ. Press. Cambridge, Mass.

Foster, D. R., and G. Motzkin. 1999. Historical influences on the landscape of Martha's Vineyard: Perspectives on the management of the Manuel F. Correllus State Forest. *Harvard Forest Paper* No. 23.

Gardner, S. C. 1997. Movements, survival, productivity, and test of a habitat suitability index model for reintroduced Columbian sharp-tailed grouse. Master's thesis, Univ. of Idaho, Moscow.

Giesen, K. M. 1987. Population characteristics and habitat use by Columbian sharp-tailed grouse in northwest Colorado. Final Report, Colorado Div. of Wildlife, Fed. Aid Project W-152-R.

———. 1994a. Breeding range and population status of lesser prairie-chickens in Colorado. *Prairie Nat.* 26:175–182.

———. 1994b. Movements and nesting habitat of lesser prairie-chicken hens in Colorado. *Southwestern Nat.* 39:96–98.

———. 1998. The lesser prairie-chicken *(Tympanuchus pallidicinctus)*. No. 364 in *The birds of North America*, A. Poole and F. Gill, eds. Academy of Natural Sciences, Philadelphia, and American Ornithologists' Union, Washington, D.C.

———. 2000. Population status and management of the lesser prairie-chicken in Colorado. *Prairie Nat.* 32:137–148.

Giesen, K. M., and I. W. Connelly. 1993. Guidelines for management of Columbian sharp-tailed grouse habitats. *Wildl. Soc. Bull.* 21:325–333.

Godfrey, W. E. 1986. *The birds of Canada*. 2nd. ed. National Museums of Canada, Ottawa, Ont.

Grange, W. B. 1948. *Wisconsin grouse problems*. Wisconsin Department of Conservation, Madison.

Gratson, M. W. 1982. Goshawks prey on radio-tagged sharp-tailed grouse. *J. Field Ornith.* 53:54–55.

———. 1988. Spatial patterns, movements, and cover selection by sharp-tailed grouse. Pp. 158–192 in *Adaptive strategies and population ecology of northern grouse*, A. T Bergerud and M. W. Gratson, eds. Univ. of Minnesota Press, Minneapolis.

———. 1989. Sexual selection on sharp-tailed grouse leks. Ph.D. diss., Univ. of Victoria, Victoria, B.C.

———. 1993. Sexual selection for increased male courtship and acoustic signals and against large male size at sharp-tailed grouse leks. *Evolution* 47:691–696.

Greenway, J. C., Jr. 1958. *Extinct and vanishing birds of the world*. Spec. Pub. No. 13, American Committee for International Wildlife Protection, New York.

Greg, L., and N. Niemuth. 2000. The history, status, and future of sharp-tailed grouse in Wisconsin. *Passenger Pigeon* 62:159–174.

Gross, A. O. 1928. The heath hen. *Mem. Boston Soc. Nat. Hist.* 6 (4): 490–587.

Gutierrez, R. J., R. M. Zink, and S. Y. Yang. 1983. Genic variation, systematic and biogeographic relationships of some Galliform birds. *Auk* 100:33–47.

Hall, F. A., and G. P. Popham. 2000. A California sage grouse study update with emphasis on nesting habitat and dispersal. Paper presented at 22nd Western States Sage and Columbian Sharp-Tailed Grouse Symposium, Redmond, Oreg., July 13–14, 2000. Abstract.

Hamerstrom, F. N., Jr. 1939. A study of Wisconsin's prairie chickens and sharp-tailed grouse. *Wilson Bull.* 51:105–120.

———. 1963. Sharptail brood habitat in Wisconsin's northern pine barrens. *J. Wildl. Manage.* 27:793–802.

Hamerstrom, F. N., Jr., and F. Hamerstrom. 1961. Status and problems of North American grouse. *Wilson Bull.* 73:284–294.

———. 1973. The prairie chicken in Wisconsin. Wisconsin Dept. Nat. Res. Tech. Bulletin 64:1–52.

Harlan, J. R. 1957. *Grasslands of Oklahoma.* Oklahoma State Univ. Press, Stillwater.

Harper, R. M. 1908. The pine barrens of Babylon and Islip, Long Island. *Torreya* 8:1–9.

Hart, C. M., O. S. Lee, and J. B. Low. 1950. *The sharp-tailed grouse in Utah: Its life history, status, and management.* Utah State Dept. of Fish and Game, Pub. No. 3.

Hartzler, J. E., and D. A. Jenni. 1988. Mate choice by female sage grouse. Pp. 240–269 in *Adaptive strategies and population ecology of northern grouse,* A. F. Bergerud and M. W Gratson, eds. Univ. of Minnesota Press, Minneapolis.

Haukos, D. A. 1988. Reproductive ecology of lesser prairie chickens. Master's thesis, Texas Tech. Univ., Lubbock.

Haukos, D. A., and G. S. Broda. 1989. Northern harrier *(Circus cyaneus)* predation of lesser prairie-chicken *(Tympanuchus pallidicinctus). J. Raptor Res.* 23:182–183.

Haukos, D. A., and L. M. Smith. 1989. Lesser prairie chicken nest site selection and vegetation characteristics in tebuthiuron-treated and untreated sand shinnery oak in Texas. *Great Basin Nat.* 49:624–626.

———. 1999. Effects of lek age on age structure and attendance of lesser prairie-chickens *(Tympanuchus pallidicinctus). Amer. Midl. Nat.* 142:415–420.

Herkert, J. R. 1991. An ecological study of the breeding birds of grassland habitats within Illinois. Ph.D. diss., Univ. of Illinois, Urbana.

Hillman, G. N., and W. W. Jackson. 1973. The sharp-tailed grouse in South Dakota. South Dakota Dept. of Game, Fish, and Parks Tech. Bulletin 3.

Hjorth, I. 1970. Reproductive behavior in the Tetraonidae, with special reference to males. *Viltrevy* 7 (4): 184–596.

Hoag, A. W., and C. E. Braun. 1990. Status and distribution of plains sharp-tailed grouse in Colorado. *Prairie Nat.* 22:97–102.

Hoffman, D. M. 1963. The lesser prairie chicken in Colorado. *J. Wildl. Manage.* 27:726–732.

Hoffman, R. W., and S. J. Silver. 2000. Current status and distribution of sage and Columbian sharp-tailed grouse. Paper presented at 22nd Western States Sage and Columbian Sharp-Tailed Grouse Symposium, Redmond, Oreg., July 13–14, 2000. Abstract.

Hoffman, R. W., W. D. Snyder, G. C. Miller, and C. E. Braun. 1992. Reintroduction of greater prairie-chickens in northeastern Colorado. *Prairie Nat.* 24:197–204.

Horak, G. J. 1985. *Kansas prairie chickens.* Kansas Fish and Game Comm., Pratt.

Horak, G. J., and R. D. Applegate. 1998. Greater prairie chicken management. *Kansas School Nat.* 45 (1): 3–16.

Horkel, J. D. 1979. Cover and space requirements of Attwater's prairie chicken *(Tympanuchus cupido attwateri)* in Refugio County, Texas. Ph.D. diss., Texas A&M Univ., College Station.

Horton, R. E. 2000. Distribution and abundance of lesser prairie-chicken in Oklahoma. *Prairie Nat.* 32:189–195.

Hupp, J. W., and C. E. Braun. 1991. Geographical variation among sage grouse populations in Colorado. *Wilson Bull.* 103:255–261.

Jackson, A. S., and R. DeArment. 1963. The lesser prairie chicken in the Texas panhandle. *J. Wildl. Manage.* 27:733–737.

Jensen, W. E., D. A. Robinson Jr., and R. D. Applegate. 2000. Distribution and population trend of lesser prairie-chicken in Kansas. *Prairie Nat.* 32:169-175.

Johnsgard, P. A. 1973. *Grouse and quails of North America.* Univ. of Nebraska Press, Lincoln.

———. 1983. *The grouse of the world.* Univ. of Nebraska Press, Lincoln.

———. 1994. *Arena birds: Sexual selection and behavior.* Smithsonian Institution Press, Washington, D.C.

———. 2000. Ecogeographic aspects of greater prairie-chicken leks in southeastern Nebraska. *Nebraska Bird Review* 68:179–184.

———. 2001. *Prairie birds: Fragile splendor in the Great Plains.* Univ. Press of Kansas, Lawrence.

Johnsgard, P. A., and R. W. Wood. 1968. Distributional changes and interactions between prairie-chickens and sharp-tailed grouse in the Midwest. *Wilson Bull.* 80:173–188.

Johnson, K. H., and C. E. Braun. 1999. Viability and conservation of an exploited sage grouse population. *Cons. Biol.* 13:77–84.

Johnson, M. D. 1964. *Feathers from the prairie.* North Dakota Game and Fish Dept., Bismarck.

Johnston, A., and S. Smoliak. 1976. Settlements of the grasslands and the greater prairie chicken. *Blue Jay* 34:153–156.

Jones, J. O. 1990. *Where the birds are: A guide to all fifty states and Canada.* William Morrow, New York.

Jones, R. E. 1963a. A comparative study of the habits of the lesser and greater prairie chickens. Ph.D. diss., Oklahoma State Univ., Stillwater.

———. 1963b. Identification and analysis of lesser and greater prairie chicken habitat. *J. Wildl. Manage.* 27:757–778.

———. 1964a. Habitat used by lesser prairie chickens for feeding related to seasonal behavior of plants in Beaver County, Oklahoma. *Southwestern Nat.* 9:111–117.

———. 1964b. The specific distinctness of the greater and lesser prairie chickens. *Auk* 81:65–73.

Keir, J. 2001. The Wisconsin prairie-chicken. Sutton Avian Res. Center, Bartlesville, Okla. *Prairie Grouse Tech. Newsletter* 3:13–14.

Kermott, L. H., and L. Oring. 1975. Acoustical communication of male sharp-tailed grouse *(Pedioecetes phasianellus)* on a North Dakota dancing ground. *Animal Behav.* 23:375–386.

Kessler, W. B., and R. P. Bosch. 1982. Sharp-tailed grouse and range management practices in western rangelands. Pp. 133–146 in *Wildlife-Livestock Relationships Symp. Pro-*

ceedings 10, J. M. Peek and P. D. Dalke, eds. Univ. of Idaho, Forest, Wildlife, and Range Exper. Station, Moscow.

Kingery, H. E., ed. 1998. *Colorado breeding bird atlas.* Colorado Breeding Bird Atlas Partnership and Colorado Div. of Wildlife, Denver.

Kirsch, L. M. 1974. Habitat management considerations for prairie chickens. *Wildl. Soc. Bull.* 2:124–129.

Kirsch, L. M., A. T. Kiett, and H. W. Miller. 1973. Land use and prairie grouse population relationships in North Dakota. *J. Wildl. Manage.* 37:449–453.

Klott, J. H., and F. G. Lindzey. 1990. Brood habitats of sympatric sage grouse and Columbian sharp-tailed grouse in Wyoming. *J. Wildl. Manage.* 54:84–88.

Kuchler, A. W. 1964. *Potential natural vegetation of the conterminous United States.* Amer. Geogr. Soc. Spec. Pub. 36.

Landel, H. 1989. A study of male and female mating behavior and female mate choice in the sharp-tailed grouse, *Tympanuchus phasianellus jamesi.* Ph.D. diss., Purdue Univ., West Lafayette, Ind.

Lehmann, V. W. 1941. *Attwater's prairie-chicken: Its life history and management.* U.S. Dept. Interior, Fish Wildl. Serv., No. Amer. Fauna 57.

———. 1968. The Attwater's prairie-chicken, current status and restoration opportunities. *Transactions of the No. Amer. Wildl. and Nat. Res. Conf.* 33:398–407.

Lehmann, V. W., and R. G. Mauermann. 1963. Status of Attwater's prairie-chicken. *J. Wildl. Manage.* 27:713–725.

Leopold, A. 1949. *A Sand County almanac and sketches from here and there.* Oxford Univ. Press, New York.

Ligon, J. S. 1927. *Wildlife of New Mexico: Its conservation and management.* New Mexico Game Comm., Santa Fe.

———. 1961. *New Mexico birds and where to find them.* Univ. of New Mexico Press, Albuquerque.

Little., E. L., Jr. 1971. *Atlas of United States trees.* Vol. 1. Misc. Pub. 1146. U.S. Dept. Agric., Forest Serv., Washington, D.C.

Litton, G., R. L. West, D. F. Dvorak, and G. T. Miller. 1994. The lesser prairie chicken and its management in Texas. Rev. ed. Booklet N7 100–025. Texas Parks and Wildl. Dept., Austin.

Locke, B. A. 1992. Lek hypothesis and the location, dispersion, and size of lesser prairie chicken leks. Ph.D. diss., New Mexico State Univ., Las Cruces.

Lucchini, V., J. Hoglund, S. Klaus, J. Swenson, and E. Randi. 2001. Historical biogeography and mitochondrial DNA phylogeny of grouse and ptarmigan. *Molecular Ecol. and Evolution* 20:149–162.

Lumsden, H. G. 1965. The displays of the sharp-tailed grouse. Ontario Dept. Lands and Forests, Tech. Ser. Res. Rep. 66.

———. 1966. The prairie-chicken in southwestern Ontario. *Can. Field-Nat.* 80:33–45.

Lutz, R. S., J. S. Lawrence, and N. J. Silvy. 1994. Nesting ecology of Attwater's prairie-chicken. *J. Wildl. Manage.* 58:230–233.

Magee, D. W., and H. E. Ahles. 1999. *Flora of the Northeast: A manual of the vascular flora of New England and adjacent New York*. Univ. of Massachusetts Press, Amherst.

Marks, J. S., and V. S. Marks. 1987. *Habitat selection by Columbian sharp-tailed grouse in west-central Idaho*. U.S. Dept. of the Interior, Bur. Land Manage., Boise District, U.S. Govt. Report 792–057/40, 019.

———. 1988. Winter habitat use by Columbian sharp-tailed grouse in western Idaho. *J. Wildl. Manage.* 52:743–746.

McClanahan, R. C. 1940. Original and present breeding ranges of certain game birds in the United States. U.S. Dept. Interior, Bur. Biol. Survey Wildl. Leaflet BS–158.

McDonald, M. W., and K. P. Reese. 1998. Landscape changes within the historical distribution of Columbian sharp-tailed grouse in eastern Washington: Is there hope? *Northwest Sci.* 72:34–41.

Mechlin, L. M. 2001. Missouri update. Sutton Avian Res. Center, Bartlesville, Okla. *Prairie Grouse Tech. Newsletter* 3:14.

Mechlin, L. M., R. W. Cannon, and D. M. Christisen. 1999. Status and management of greater prairie chickens in Missouri. Pp. 129–142 in The greater prairie chicken: A national look, W. D. Svedarsky, R. H. Hier, and N. J. Silvy, eds. Univ. of Minnesota, St. Paul, Misc. Pub. 9–1999.

Meints, D. R. 1991. Seasonal movements, habitat use, and productivity of Columbian sharp-tailed grouse in southeastern Idaho. Master's thesis, Univ. of Idaho, Moscow.

Meints, D. R., J. W. Connelly, K. P. Reese, A. R. Sands, and T. P. Hemker. 1992. Habitat suitability index procedure for Columbian sharp-tailed grouse. Univ. of Idaho, Forest, Wildlife, and Range Exper. Station Bulletin 55.

Merchant, S. S. 1982. Habitat-use, reproductive success, and survival of female lesser prairie chickens in two years of contrasting weather. Master's thesis, New Mexico State Univ., Las Cruces.

Merrill, M. D., K. A. Chapman, and K. A. Poiani. 1999. Land-use patterns surrounding greater prairie-chicken leks in northwestern Minnesota. *J. Wildl. Manage.* 63:189–198.

Miller, G. C., and W. D. Graul. 1980. Status of sharp-tailed grouse in North America. Pp. 18–28 in *Proceedings of the Prairie Grouse Symposium*, P. A. Vohs and F. L. Knopf, eds. Oklahoma State Univ., Stillwater.

Moe, M. 1999. Status and management of the greater prairie chicken in Iowa. Pp. 123–127 in The greater prairie chicken: A national look, W. D. Svedarsky, R. H. Hier, and N. J. Silvy, eds. Univ. of Minnesota, St. Paul, Misc. Pub. 9–1999.

Morrow, M. E. 1986. Ecology of Attwater's prairie chicken in relation to land management practices on the Attwater Prairie Chicken National Wildlife Refuge. Ph.D. diss., Texas A&M Univ., College Station.

Moseley, R., and C. Groves. 1990. *Rare, threatened, and endangered plants and animals of Idaho*. Natural Heritage Section, Nongame and Endangered Wildlife Program, Idaho Dept. of Fish and Game, Boise.

Mote, K. D., R. D. Applegate, J. A. Bailey, K. E. Giesen, R. Horton, and J. L. Sheppard, eds. 1999. *Assessment and conservation strategy for the lesser prairie chicken* (Tympanuchus pallidicinctus). Kansas Dept. of Wildlife and Parks, Emporia.

Moyles, D. L. J., and D. A. Boag. 1981. Where, when, and how male sharp-tailed grouse establish territories on arenas. *Can. J. Zool.* 59:1576–1581.

Niemuth, N. D. 2000. Land use and vegetation associated with greater prairie-chicken leks in an agricultural landscape. *J. Wildl. Manage.* 64:278–286.

Niering, W. A. 1953. The past and present vegetation of High Point State Park, New Jersey. *Ecol. Monogr.* 23:127–148.

Nyland, P. 2001. Status of sage-grouse at the U.S. Army Yakima Training Center, Washington. *Prairie Grouse Tech. Newsletter* 3:17.

Oberholser, H. C. 1974. *The bird life of Texas.* Vol. 1. Univ. of Texas Press, Austin.

Oedekoven, O. O. 1985. Columbian sharp-tailed grouse population distribution and habitat use in south central Wyoming. Master's thesis, Univ. of Wyoming, Laramie.

Ogden, J. G., III. 1961. Forest history of Martha's Vineyard, Massachusetts. 1. Modern and pre-colonial forests. *Amer. Midl. Nat.* 66:417–431.

Olawsky, G. D., and L. M. Smith. 1991. Lesser prairie-chicken densities on tebuthiuron-treated and untreated sand shinnery oak rangelands. *J. Range Manage.* 44:364–368.

Olsvig, L. S., J. F. Cryan, and R. H. Whittaker. 1979. Vegetational gradients of the pine plains and barrens of Long Island, New York. Pp. 265–282 in *Pine barrens: Ecosystem and landscape,* R. T. Forman, ed. Academic Press, New York.

Oyler-McChance, S. J., K. P. Burnham, and C. E. Braun. 2001. Influence of changes in sagebrush on Gunnison sage-grouse in southwestern Colorado. *Southwest. Nat.* 46:323–331.

Parker, T. L. 1970. On the ecology of the sharp-tailed grouse in southeastern Idaho. Master's thesis, Idaho State Univ., Boise.

Partch, M. 1973. A history of Minnesota's prairie chickens. Pp. 15–29 in *The prairie chicken in Minnesota,* W. D. Svedarsky and T. J. Wolfe, eds. Conference proceedings, April 28, 1973, Univ. of Minnesota, Crookston.

Peterson, M. J. 1994. Factors limiting population size of the endangered Attwater's prairie chicken *(Tympanuchus cupido).* Ph.D. diss., Texas A&M Univ., College Station.

Peterson, M. J., and N. J. Silvy. 1996. Reproductive stages limiting productivity of the endangered Attwater's prairie chicken. *Conserv. Biol.* 10:1264–1276.

Peterson, R. S., and C. S. Boyd. 1998. Ecology and management of sand shinnery communities: A literature review. U.S. Dept. Agric., Forest Service, Gen. Tech. Rep. RMRS-GTR-16.

Phillips, J. B. 1990. Lek behaviour in birds: Do displaying males reduce nest predation? *Anim. Behav.* 39:555–565.

Principal Game Birds and Mammals of Texas. 1945. Texas Game, Fish, and Oyster Comm., Austin.

Prose, B. L. 1987. Habitat suitability index models: Plains sharp-tailed grouse. U.S. Fish Wildl. Serv. Biol. Rep. 82 (10.142).

Pyne, S. 1982. *Fire in America: A cultural history of wildland and rural fire.* Princeton Univ. Press, Princeton.

Rakestraw, J. 1995. A closer look: Lesser prairie-chicken. *Birding* 27 (6): 109–112.

Remington, T. E., and C. E. Braun. 1985. Sage grouse food selection in winter, North Park, Colorado. *J. Wildl. Manage.* 49:1055–1061.

Riley, T. Z. 1978. Nesting and brood rearing habitat of lesser prairie chickens. Master's thesis, New Mexico State Univ., Las Cruces.

Riley, T. Z., C. A. Davis, M. Ortiz, and M. J. Wisdom. 1992. Vegetative characteristics of successful and unsuccessful nests of lesser prairie-chickens. *J. Wildl. Manage.* 56:383–387.

Riley, T. Z., C. A. Davis, and R. A. Smith. 1993. Autumn and winter foods of the lesser prairie-chicken *(Tympanuchus pallidicinctus)* (Galliformes: Tetraonidae). *Great Basin Nat.* 53:186–189.

Ritcey, R. 1995. *Status of the sharp-tailed grouse in British Columbia.* Wildlife Working Report No. WR–70. Ministry of Environment–Wildlife Branch, Victoria, B.C.

Robbins, C. S., and E. A. T. Blum, eds. 1996. *Atlas of the breeding birds of Maryland and the District of Columbia.* Univ. of Pittsburgh Press, Pittsburgh.

Robel, R. J. 1966. Booming territory size and mating success of the greater prairie chicken *(Tympanuchus cupido pinnatus). Anim. Behav.* 14:328–331.

———. 1967. Significance of booming grounds of the greater prairie chicken. *Proceedings of the Amer. Philos. Soc.* 111:109–114.

———. 1970. Possible role of behavior in regulating greater prairie chicken populations. *J. Wildl. Manage.* 34:306–312.

Robel, R. J., and W. B. Ballard Jr. 1974. Lek social organization and reproductive success in the greater prairie chicken. *Amer. Zool.* 14:121–128.

Robel, R. J., J. N. Briggs, J. J. Cebula, N. J. Silvy, C. E. Viers, and P. G. Watt. 1970. Greater prairie chicken ranges, movements, and habitats in Kansas. *J. Wildl. Manage.* 34:286–306.

Robel, R. J., F. R. Henderson, and W. Jackson. 1972. Some sharp-tailed grouse population statistics from South Dakota. *J. Wildl. Manage.* 36:87–98.

Robichaud, B., and M. F. Buell. 1973. *Vegetation of New Jersey: A study of landscape diversity.* Rutgers Univ. Press, New Brunswick, N.J.

Rodgers, R. 2001. Kansas prairie grouse survey results, 2001. Sutton Avian Res. Center, Bartlesville, Okla. *Prairie Grouse Tech. Newsletter* 3:14–15.

Rodgers, R. D, and M. L. Sexson. 1990. Impacts of extensive chemical control of sand sagebrush on breeding birds. *J. Soil and Water Conserv.* 45:494–497.

Rogers, G. E. 1969. The sharp-tailed grouse in Colorado. Colorado Game, Fish, and Parks Dept. Tech. Pub. 23.

Ryan, M. L., W. Burger Jr., and D. P. Jones. 1998. Breeding ecology of greater prairie-chickens *(Tympanuchus cupido)* in relation to prairie landscape configuration. *Amer. Midl. Nat.* 140:111–121.

Saab, V. A., and J. S. Marks. 1992. Summer habitat use by Columbian sharp-tailed grouse in western Idaho. *Great Basin Nat.* 52:166–173.

Sands, J. L. 1968. Status of the lesser prairie-chicken. *Audubon Field Notes* 22:454–456.

Schroeder, M. A. 1991. Movement and lek visitation by female greater prairie-chickens in relation to predictions of the female-preference hypothesis of lek evolution. *Auk* 108:896–903.

———, 2000a. Dispersion of nests in relation to lek locations for sage grouse in north-central Washington. Paper presented at 22nd Western States Sage and Columbian Sharp-Tailed Grouse Symposium, Redmond, Oreg., July 13–14, 2000. Abstract.

———. 2000b. Distribution, abundance, and management of Columbian sharp-tailed grouse in Washington. *P. G. News* (Prairie Grouse Technical Council Newsletter 2), November, p. 9.

———. 2001. Distribution and abundance of Columbian sharp-tailed grouse in Washington. *P. G. News* (Prairie Grouse Technical Council Newsletter 3), November, pp. 14–15.

Schroeder, M. A., D. W. Hays, M. A. Murphy, and D. J. Pierce. 2000. Changes in the distribution and abundance of Columbian sharp-tailed grouse in Washington. *Northwestern Nat.* 81:95–103.

Schroeder, M. A., and L. A. Robb. 1993. Greater prairie-chicken *(Tympanuchus cupido).* No. 36 in *The birds of North America,* A. Poole and F. Gill, eds. Academy of Natural Sciences, Philadelphia, and American Ornithologists' Union, Washington, D.C.

Schroeder, M. A., J. R. Young, and C. E. Braun. 1999. Sage grouse *(Centrocercus urophasianus).* No. 425 in *The birds of North America,* A. Poole and F. Gill, eds. Academy of Natural Sciences, Philadelphia, and American Ornithologists' Union, Washington, D.C.

Schwartz, C. W. 1945. The ecology of the prairie chicken in Missouri. *Univ. of Missouri Studies* 20:1–99.

Sedivec, K. K. 1994. Grazing treatment effects on and habitat use of upland nesting birds on native grassland. Ph.D. diss., North Dakota State Univ., Fargo.

Sell, D. L. 1979. Spring and summer movements and habitat use by lesser prairie chicken females in Yoakum County, Texas. Master's thesis, Texas Tech. Univ., Lubbock.

Seyffert, K. D. 2001. *Birds of the Texas panhandle: Their status, distribution, and history.* Texas A&M Univ. Press, College Station.

Sharpe, R. S. 1968. The evolutionary relationships and comparative behavior of prairie chickens. Ph.D. diss., Univ. of Nebraska, Lincoln.

Shiller R. J. 1973. Reproductive ecology of female sharp-tailed grouse and its relation to early plant succession in NW Minnesota. Ph.D. diss., Univ. of Minnesota, Minneapolis.

Short, L. L. 1967. A review of the genera of grouse (Aves, Tetraoninae). *Amer. Mus. Novit.* no. 2289, pp. 1–39.

Sisson, L. 1976. The sharp-tailed grouse in Nebraska. Nebraska Game and Parks Comm., Research report W–38–R.

Snyder, J. W., E. C. Pilren, and J. A. Crawford. 1999. Translocation histories of prairie grouse in the United States. *Wildl. Soc. Bull.* 27 (2): 428–432.

Snyder, W. A. 1967. Lesser prairie chicken. Pp. 121–128 in *New Mexico wildlife management.* New Mexico Dept. Game and Fish, Sante Fe.

Sparling, D. W., Jr. 1979. Reproductive isolating mechanisms and communication in greater prairie chickens *(Tympanuchus cupido)* and sharp-tailed grouse *(Tympanuchus phasianellus).* Ph.D. diss., Univ. of North Dakota, Grand Forks.

———. 1980. Hybridization and taxonomic status of greater prairie-chickens and sharp-tailed grouse. *Prairie Nat.* 12:92–101.
———. 1981. Communication in prairie grouse. 2. Ethological isolating mechanisms. *Behav. Neural Biol.* 32:487–503.
———. 1983. Quantitative analysis of prairie grouse vocalizations. *Condor* 5:30–42.
Stempel, M. E., and S. Rogers Jr. 1961. History of prairie chickens in Iowa. *Proceedings of the Iowa Acad. Sci.* 68:314–322.
Sullivan, R. M., J. P. Hughes, and J. E. Lionberger. 2000. Review of the historical and present status of the lesser prairie-chicken *(Tympanuchus pallidicinctus)* in Texas. *Prairie Nat.* 32:177–188.
Suminski, H. R. 1977. Habitat evaluation for lesser prairie chickens in eastern Chaves County, New Mexico. Master's thesis, New Mexico State Univ., Las Cruces.
Sutton, G. M. 1967. *Oklahoma birds.* Univ. of Oklahoma Press, Norman.
———. 1977. The lesser prairie chicken's inflatable neck sacs. *Wilson Bull.* 89:521–522.
Svedarsky, W. D. 1979. Spring and summer ecology of greater prairie chickens in northwestern Minnesota. Ph.D. diss., Univ. of North Dakota, Grand Forks.
———. 1988. Reproductive ecology of female greater prairie chickens in Minnesota. Pp. 193–239 in *Adaptive strategies and population ecology of northern grouse,* A. T. Bergerud and M. W. Gratson, eds. Univ. of Minnesota Press, Minneapolis.
Svedarsky, W. D., T. J. Wolfe, and J. E. Toepfer. 1997. The greater prairie-chicken in Minnesota. *Minn. Wildl. Rep.* No. 11:1–19.
Svedarsky, W. D., R. H. Hier, and N. J. Silvy, eds. 1999. The greater prairie-chicken: A national look. Univ. of Minnesota, St. Paul, Misc. Pub. 9–1999.
Svedarsky, W. D., and T. J. Wolfe, eds. 1973. *The prairie chicken in Minnesota.* Conference Proceedings, April 28, 1973. Univ. of Minnesota, Crookston.
Sveum, C. M., J. A. Crawford, and W. D. Edge. 1998. Use and selection of brood-rearing habitat by sage-grouse in south central Washington. *Great Basin Nat.* 58:344–351.
Swenson, J. E. 1985. Seasonal habitat use by sharp-tailed grouse, *Tympanuchus phasianellus,* on mixed-grass prairie in Montana. *Can. Field-Nat.* 99:40–46.
Taylor, M. A. 1979. Lesser prairie chicken use of man-made leks. *Southwestern Nat.* 24:683–714.
Taylor, M. A., and F. S. Guthery. 1980. Status, ecology, and management of the lesser prairie chicken. U.S. Dept. Agric., Forest Serv., Gen. Tech. Rep. RM–77.
Tharp, B. C. 1906. Structure of Texas vegetation east of the 98th meridian *Univ. Tex. Bull.* 2606.
Thompson, M. C., and C. Ely. 1989. *Birds in Kansas.* Vol. 1. Univ. Press of Kansas, Lawrence.
Tirhi, M. J. 1995. *Washington state management plan for sharp-tailed grouse.* Wildlife Management Prog., Washington Department of Fish and Wildlife, Olympia.
Toepfer, J. E., R. L. Eng, and R. K. Anderson. 1990. Translocating prairie grouse: What have we learned? *Transactions of the No. Amer. Wildl. and Nat. Res. Conf.* 55:569–579.
Tsuji, L. J. S., D. R. Kozlovic, and M. B. Sokolowski. 1992. Territorial position in sharp-tailed grouse leks: The probability of fertilization. *Condor* 94:1030–1031.

Bibliography

Tsuji, L. J. S., D. R. Kozlovic, M. B. Sokolowski, and R. I. C. Hansell. 1994. Relationship of body size of male sharp-tailed grouse to location of individual territories on leks. *Wilson Bull.* 106:329–337.

Tweit, S. 2000. The next spotted owl? *Audubon* 102 (6): 64–72.

Ulliman, M. J. 1995. Winter habitat ecology of Columbian sharp-tailed grouse in southeastern Idaho. Master's thesis, Univ. of Idaho, Moscow.

U.S. Department of the Interior. 1989. Endangered and threatened wildlife and plants: Annual notice of review; proposed rules. *Federal Register* 54:560.

———. U.S. Fish and Wildlife Service. 1999. Endangered and threatened wildlife and plants: 90- day finding on a petition to list the Columbian sharp-tailed grouse as threatened. *Federal Register* 64, no. 206, pp. 57620–57623.

Vance, D. R., and R. L. Westemeier. 1979. Interactions of pheasants and greater prairie chickens in Illinois. *Wildl. Soc. Bull.* 7:221–225.

Van Dyke, W., G. P. Keister, and C. E. Braun. 2000. Population characteristics of northern sage grouse in Oregon and Colorado. Paper presented at 22nd Western States Sage and Columbian Sharp-Tailed Grouse Symposium, Redmond, Oreg., July 13–14, 2000. Abstract.

Wachob, D. C. 1997. The effects of the Conservation Reserve Program on wildlife in southeastern Wyoming. Ph.D. diss., Univ. of Wyoming, Laramie.

Wakkinen, W. L., K. P. Reese, and J. W. Connelly. 1992. Sage-grouse nest locations in relation to leks. *J. Wildl. Manage.* 56:381–383.

Walker, L., ed. 2000. Proceedings of the 22nd Western States Sage and Columbian Sharp-Tailed Grouse Symposium. Redmond, Oreg., July 13–14, 2000. (Abstracts available online: http://rangenet.org/projects/grouse/22ndabstracts.html.)

Weidensaul, S. 2001. Sage grouse strut their stuff. *Smithsonian* 33 (3): 56–63.

Westemeier, R. L., and S. Gough. 1999. National outlook and conservation needs for greater prairie chickens. Pp. 169–187 *in* The greater prairie chicken: A national look, W. D. Svedarsky, R. H. Hier, and N. J. Silvy, eds. Univ. of Minnesota, St. Paul, Misc. Pub. 9–1999.

Westemeier, R. L., S. A. Simpson, and T. L. Esker. 1999. Status and management of greater prairie chickens in Illinois. Pp. 143–152 *in* The greater prairie chicken: A national look, W. D. Svedarsky, R. H. Hier, and N. J. Silvy, eds. Univ. of Minnesota, St. Paul, Misc. Pub. 9–1999.

Whitney, G. G. 1994. *From coastal wilderness to fruited plain*. Cambridge Univ. Press, Cambridge, Eng.

Wiedeman, V. E., and W. T. Penfound. 1960. A preliminary study of the shinnery in Oklahoma. *Southwestern Nat.* 5:117–122.

Wilcove, D. 1999. *The condor's shadow: The loss and recovery of wildlife in America*. W. H. Freeman, New York.

Wiley, R. H. 1978. The lek mating system of the sage grouse. *Sci. Amer.* 238 (5): 114–125.

Wisdom, M. J. 1980. Nesting habitat of lesser prairie chickens in eastern New Mexico. Master's thesis, Univ. of New Mexico, Las Cruces.

Wood, D. S., and G. D. Schnell. 1984. *Distribution of Oklahoma birds*. Univ. of Oklahoma Press, Norman.

Woodward, A. J., S. D. Fuhlendorph, D. M. Leslie Jr., and J. Shackford. 2001. Influence of landscape composition and change on lesser prairie-chicken *(Tympanuchus pallidicinctus)* populations. *Amer. Midl. Nat.* 145:261–274.

Yeatter, R. E. 1943. The prairie chicken in Illinois. *Ill. Nat. Hist. Survey Bull.* 22:377–416.

———. 1963. Population responses of prairie chickens to land-use changes in Illinois. *J. Wildl. Manage.* 27:739–757.

Yost, J. A. 2001. Colorado lesser prairie-chicken breeding survey, 2001. Sutton Avian Res. Center, Bartlesville, Okla. *Prairie Grouse Tech. Newsletter* 3:15–16.

Young, D. E. 1953. Ecological considerations in the extinction of the passenger pigeon *(Ectopistes migratorius)*, heath hen *(Tympanuchus cupido cupido)*, and the Eskimo curlew *(Numenius borealis)*. Ph.D. diss., Univ. of Michigan, Ann Arbor.

Young, J. R., C. E. Braun, S. J. Oyler-McCance, J. W. Hupp, and T. W. Quinn. 2000. A new species of sage grouse from southwestern Colorado. *Wilson Bull.* 112:445–453.

Young, J. R., J. W. Hupp, J. W. Bradbury, and C. E. Braun. 1994. Phenotypic divergence of secondary sexual traits among sage grouse, *Centrocercus urophasianus*, populations. *Anim. Behav.* 47:1353–1362.

INDEX

This index includes major geographic locations mentioned in the text, authorities cited, and grouse taxa and other bird species (primarily indexed by vernacular names), but it excludes most other plants and animals except for some significant species. The major descriptive account for each grouse taxon is indicated in **bold,** and maps or drawings are indicated in *italics*.

Alaska, 92
Alberta, Canada, 53, 90
Aldrich, John, 4
Aldridge, Cameron, 110
American crow, 16
American Lands Alliance, 122
American Ornithologists' Union (AOU), 121
Apa, Anthony, 102
Applegate, Roger, 10, 66
Aransas National Wildlife Refuge, 26
Arizona, 3, 108, 121
Arkansas, 53
Arkansas River, 40, 41, 127
aromatic sumac, 29
Askins, Robert, 5, 8
aspens, 95
Attwater, H. P., 18
Attwater Prairie Chicken National Wildlife Refuge, 26, 27, 28, 129

Attwater's prairie-chicken, 4, 13, **17–28,** *19*, *24*, 32, 51, 68, 72, 79, 128, 129, 136
Audubon, John J., 3

Bailey, James, 36
Bain, Mathew, 40
Baird's sparrow, 128
Baker, Maurice, 66
Ballard, Warren, 71
Baydack, Richard, 95, 97
bear oak, 6, 7, 9
Bent, Arthur C, 10, 54
Berger, Robert, 95
big bluestem, *54*, 104, 132
big sagebrush, 105, 108, *109*, 121. *See also* sagebrush
Biodiversity Legal Foundation, 91, 94, 117
birches, 95
black-billed magpie, 5
blackhead (disease), 2, 16
blackjack oak, 66, 67
Black Kettle National Grassland, 127
black-throated sparrow, 127
blueberry, 1, 4, 6, 7, 9
blue grama, 132
blue grouse, 104
bobcat, 20
Boisvert, Jennifer, 103
Bonaparte, Charles, 105
Bonaparte, Napoleon, 105

Index

Boyd, Chard, 127
Braun, Clait, 93, 108, 110, 116, 121
British Columbia, Canada, 82, 90, 91
brown-headed cowbird, 5
Bry, Ed, 96
Buell, M., 6
Buena Vista Marsh, 59, 69
buffaloberry, 95
buffalo grass, 42
Burchard Lake, 132, 133
Bureau of Land Management (BLM), 31, 38, 50, 122, 126, 127
burrowing owl, 5, 128

California, 82, 91, 108
Candidate Species Conservation Agreements, 50
capercaillie, 105, 107, 112, 114
Carolina parakeet, 16
Cartright, Kelly, 71
Cather, Willa, 80
Centrocercus, 135
   minimus, 121, 135
   urophasianus, 135
Chihuahuan raven, 127
Cimarron National Grassland, 41, 127
Cimarron River, 40
Colorado, 32, 34, 41, 49, 51, 62, 65, 68, 82, 85, 87, 90, 91, 102, 103, 118, 121, 127, 128, 132
Columbian sharp-tailed grouse, 82, 88, **91–94**, 92
Colville Indian Reservation, 93
Comanche National Grasslands, 41, 127
Connecticut River, 3, 8
Connelly, J. W., 130
Conservation Reserve Program (CRP), 35, 41, 42, 62, 64, 66, 88, 90, 103, 131
Copelin, Farrell, 43–49
cottonwood, 95
coyote, 20, 40, 79
Crawford, John, 44

Day, Gordon, 8
Dendragapus, 104
Dimare, Maria, 23
DNA, 17
downy brome, 132

Eagle Lake, 26
eastern meadowlark, 17
Edwards Plateau, 29
Ely, Charles, 41
Endangered Species Act, 26, 78, 94, 110, 122
Evans, R., 97

Farney, Dennis, 124
fire ant, 20
Flint Hills (Kans.), 65, 67, 133
Fossil Rim Wildlife Center, 26

Galveston Bay Prairie Preserve, 27
Giesen, Kenneth, 41, 42, 47
Gough, Sharon, 63, 64, 66, 67, 128
grama grass, 29, 42
grasshopper sparrow, 5, 8, 10, 128
Gratson, M. W., 97
Graul, Walter, 84, 90, 91, 93, 94
greater prairie-chicken, 5, 16, 17, 21, 42, 44, 49, **52–80,** 81, 95, 96, 126, 128
greater sage-grouse, **104–118,** 106, 109, 113, 114, 115, 119, 120, 122, 129, 130, 135
great horned owl, 7
Gross, Alfred O., 10, 12, 14, 15, 16, 59, 72
Gunnison sage-grouse, 104, 108, 109, **118–122,** 120, 121, 125, 136

hairy grama (grass), 33
Hamerstrom, Frances, 59, 69, 70, 71, 75, 78
Hamerstrom, Fred, 59, 69, 70, 71, 75, 78
Hartzler, J., 111
heath hen, **1–16,** 9, 11, 13, 22, 28, 32, 129, 136
Hiawatha National Forest, 85
Hjorth, Ingemar, 44–46, 48, 70, 72–75, 77, 97–102, 114, 115
Hoffman, Richard, 91, 93
Horak, Gerald, 66, 71
Horkel, John, 23, 25
horned lark, 17
Horton, R., 39
huckleberry, 1, 4, 6, 7
Hupp, Jerry, 118
hybrid grouse, 27, 32, 40, 44, 96, 104, 114, 116

Idaho, 91, 93, 108
Illinois, 16, 60, 129
Indian grass, 17, 132
interior greater prairie-chicken, 4, 18, 21, 23, 32, 34, 40, **52–80,** 54, 72–77, 136. *See also* greater prairie-chicken
Iowa, 54, 61, 82, 84, 111

Jensen, William, 41
Jones, John, 126

Kansas, 32, 34, 38, 40, 41, 51, 61–63, 65, 66, 68, 82, 87, 108, 121, 127, 128, 133
Kansas–Illinoian glaciation, 6
Kentucky, 53

154

Index

Kermott, Henry, 99
Konza Prairie, 133
Kuchler, A. W., 125

lark bunting, 128
lark sparrow, 128
Lehmann, Valgene W., 18, 19
Leif, Tony, 89
Leopold, Aldo, 59, 68, 85
lesser prairie-chicken, 17, 21, **29–51**, *33*, *45–47*, 67, 68, 73, 79, 127
Lesser Prairie-Chicken Interstate Working Group, 50
Lewis and Clark expedition, 105
little bluestem, 7, 23, 25, 30, *33*,132
Locke, Brian, 47, 48
loggerhead shrike, 127
long-billed curlew, 128
Long Island, 3, 6, 7, 8
Louisiana, 20
Lucchini, V., 104
Lumsden, Harry, 97, 98, 100
Lutz, R., 25

Maine, 6, 7, 10
Manitoba, Canada, 82, 83, 90, 95, 96
Manitoulin Island, Canada, 84, 95, 96
maple, 95
marsh wren, 17
Martha's Vineyard, Mass., 1–16, 129
Martha's Vineyard State Forest, 1
Maryland, 3
McCown's longspur, 128
Meade Wildlife Area, 59
Mechlin, Larry, 62
Michigan, 55, 59, 60, 82, 84, 85
Michigan Department of Natural Resources, 85
Miller, Gary, 84, 90, 91, 93, 94
Minnesota, 54, 57, 61, 68, 84, 128
Minnesota Prairie Chicken Preserve System, 58
Minnesota Prairie Chicken Society, 57
Missouri, 61, 68
Missouri Conservation Department, 61, 62
Missouri Prairie Foundation, 62
Moe, Mel, 61
Montana, 82, 87, 88, 90, 91, 93, 108
Morrow, Michael, 25
Muleshoe National Wildlife Refuge, 127

National Forest Service, 126
National Park Service, 126

Nature Conservancy, 26, 27, 58, 60, 124
Nebraska, 7, 32, 52, 61–64, 67, 68, 88, 90, 95, 96, 103, 108, 123, 131, 132
Nebraska Game and Parks Commission, 64
needle-and-thread (grass), 29, *33*, 132
Nevada, 108
New Hampshire, 7
New Jersey, 3, 5, 6, 7
New Mexico, 29, 31, 32, 34, 36, 38, 39, 50, 51, 82, 108, 121
New Mexico Cattle Growers Association, 38
New Mexico Department of Game and Fish, 31, 38
New Mexico Heritage Institute, 36
New York, 3, 7, 8
Nine-Mile Prairie (Nebr.), 123–124
North Carolina, 4, 5
North Dakota, 54–56, 82, 87, 89, 90, 95, 101, 108
northern goshawk, 16
northern greater sage-grouse, **104–118**. *See also* sage-grouse
northern harrier, 16

Ogallala aquifer, 31, 34, 35
Oglala National Grasslands, 88
Ohio, 53
Oklahoma, 31, 32, 34, 38–40, 43, 50, 51, 53, 62, 66–68, 82, 87, 108, 121, 127
Olsvig, Linda, 7
Ontario, 82, 83
opossum, 20
Optima National Wildlife Refuge, 127

Packsaddle Wildlife Management Area, 127
Partch, Max, 57
partridge berry, 8, 9
paspalum (grass), 21, 23, 132
passenger pigeon, 16
Pennsylvania, 3, 5, 7, 8, 10
Peterson, Markus, 25
Peterson, Roger, 127
Philadelphia Academy of Arts and Sciences, 105
pinnated grouse, 32, 135
pitch pine, 7
plains sharp-tailed grouse, 82, **85–95**, *86*, 95, 103
Pleistocene geologic period, 4, 5, 17, 18, 32
Pocono Plateau, 3, 10
poplar, 87
post oak, 38
Potomac River, 4

Index

prairie-chicken. *See* greater prairie-chicken; lesser prairie-chicken
Prairie Chicken Foundation of Illinois, 60
Prairie Chicken Foundation (Wis.), 59
Prairie Ridge State Natural Area, 60
prairie sharp-tailed grouse, **82–85,** *83,* 95, 125
Pyne, Stephen, 8

quaking aspen, *83*
Quivera National Wildlife Refuge, 40

rabbitbrush, 95
raccoon, 20
red-bellied woodpecker, 7
Red River Valley (N.Dak.), 55–58, 87–90
red-winged blackbird, 17
red wolf, 20
Reese, K. P., 130
Remington, T., 110
reticuloendothelial virus (REV), 27
Rhode Island, 7
Ringgold Wildlife Area, 61
Robel, Robert, 71

sagebrush, 88, 95, 105–108, 110, 125, 129. *See also* big sagebrush; sand sagebrush
sage grouse. *See* greater sage-grouse; Gunnison sage-grouse
sage sparrow, 118
sage thrasher, 118
sand dropseed (grass), 29, 30, 37, 132
Sandhills (Nebr.), 64, 87, 95
sand lovegrass, 29
sand sagebrush, 30, 37, 39, 43, 129
Sartore, Joel, 28
Saskatchewan, 53, 83, 90, 110
Sauer, John, 90
Savannah sparrow, 5, 6
scaled quail, 127
Schmidt, F. G., 59
Schroeder, Michael, 70, 94
Schwartz, Charles, 61, 62
seaside sparrow, 17
serviceberry, 86, 87, 95
Seyffert, Kenneth, 35
Sharpe, Roger, 23, 24, 44, 45, 48
sharp-tailed grouse, *5,* 17, 18, 30, 32, 45, 57, 62, 63, 65, 72, **81–103,** *83, 86, 92, 98, 100, 101, 102, 124, 128,* 129, 136. *See also* Columbian sharp-tailed grouse; plains sharp-tailed grouse; prairie sharp-tailed grouse
Sheyenne National Grasslands, 56
shinnery oak, 30, *31,* 39, 50

side-oats grama, 86, 130
Silver, San Juan, 91, 93
Silvy, Nova, 25
Sisson, Leonard, 88
skunks, 20, 50
snakes, 20, 50
snowberry, 95
soapweed, 29
Society of Tympanuchus Cupido Pinnatus, 59
South Dakota, 30, 62, 63, 68, 87–89, 108
Strunk, W., 116
Sullivan, Robert, 34
Svedarsky, Dan, 76
Swanson Lake Wildlife Area, 93
switchgrass, 17, 132

Tallgrass Prairie Preserve, 67
tebuthiuron, 31
Tennessee, 53
Texas, 17–41, 51, 53, 82, 87
Texas Game and Parks Department, 35
Texas Game, Fish, and Oyster Commission, 32
Thompson, Max, 41
three-awn grass, 29
Tsuji, Leonard, 97
Turner, Ted, 124
Tveten, John, 13
Tweit, Susan, 117
Tympanuchus
  cupido, 136
    attwateri, 136
    cupido, 136
    pinnatus, 136
  lulli, 5
  pallidicinctus, 136
  phasianellus, 135
    campestris, 82
    columbianus, 82
    hueyi, 82
    jamesi, 82

upland sandpiper, 8, 10
U.S. Fish and Wildlife Service, 19, 50, 91, 110, 117, 122, 126
Utah, 30, 91, 122

vesper sparrow, 8, 10
Virginia, 3, 4

Wachob, Douglas, 102
Washington, D.C., 3
Washington State, 91, 93, 94, 103, 108, 110, 117
Washita National Wildlife Refuge, 127

Index

Weaver, John, 80
Westemeier, Ronald, 63, 64, 66, 67, 128
western greater sage-grouse, 108
western wheatgrass, 92
Wilcove, David, 124
wild plum, 29, 38
wild rose, *83*, 87
Wiley, R. Haven, 111

Williams, Sartor O., 36
willow, 87, 95
Wisconsin, 54, 55, 58, 59, 68, 82, 84, 85, 95, 128
Wisconsinian glacial period, 10, 18
World Conservation Union, 122
Wyoming, 82, 87, 88, 90, 93, 103, 108

Young, Jessica, 118, 121